Solar Cell
Technology and
Applications

OTHER AUERBACH PUBLICATIONS

Advances in Semantic Media Adaptation and Personalization, Volume 2
Marios Angelides
ISBN: 978-1-4200-7664-6

Architecting Secure Software Systems
Manish Chaitanya and Asoke Talukder
ISBN: 978-1-4200-8784-0

Architecting Software Intensive Systems: A Practitioners Guide
Anthony Lattanze
ISBN: 978-1-4200-4569-7

Business Resumption Planning, Second Edition
Leo Wrobel
ISBN: 978-0-8493-1459-9

Converging NGN Wireline and Mobile 3G Networks with IMS: Converging NGN and 3G Mobile
Rebecca Copeland
ISBN: 978-0-8493-9250-4

Delivering Successful Projects with TSP^{SM} and Six Sigma: A Practical Guide to Implementing Team Software Process^{SM}
Mukesh Jain
ISBN: 978-1-4200-6143-7

Designing Complex Systems: Foundations of Design in the Functional Domain
Erik Aslaksen
ISBN: 978-1-4200-8753-6

The Effective CIO: How to Achieve Outstanding Success through Strategic Alignment, Financial Management, and IT Governance
Eric Brown and William Yarberry, Jr.
ISBN: 978-1-4200-6460-5

Enterprise Systems Backup and Recovery: A Corporate Insurance Policy
Preston Guise
ISBN: 978-1-4200-7639-4

Essential Software Testing: A Use-Case Approach
Greg Fournier
ISBN: 978-1-4200-8981-3

The Green and Virtual Data Center
Greg Schulz
ISBN: 978-1-4200-8666-9

How to Complete a Risk Assessment in 5 Days or Less
Thomas Peltier
ISBN: 978-1-4200-6275-5

HOWTO Secure and Audit Oracle 10g and 11g
Ron Ben-Natan
ISBN: 978-1-4200-8412-2

Information Security Management Metrics: A Definitive Guide to Effective Security Monitoring and Measurement
W. Krag Brotby
ISBN: 978-1-4200-5285-5

Information Technology Control and Audit, Third Edition
Sandra Senft and Frederick Gallegos
ISBN: 978-1-4200-6550-3

Introduction to Communications Technologies: A Guide for Non-Engineers, Second Edition
Stephan Jones, Ron Kovac, and Frank M. Groom
ISBN: 978-1-4200-4684-7

IT Auditing and Sarbanes-Oxley Compliance: Key Strategies for Business Improvement
Dimitris Chorafas
ISBN: 978-1-4200-8617-1

The Method Framework for Engineering System Architectures
Peter Capell, DeWitt T. Latimer IV, Charles Hammons, Donald Firesmith, Tom Merendino, and Dietrich Falkenthal
ISBN: 978-1-4200-8575-4

Network Design for IP Convergence
Yezid Donoso
ISBN: 978-1-4200-6750-7

Profiling Hackers: The Science of Criminal Profiling as Applied to the World of Hacking
Raoul Chiesa, Stefania Ducci, and Silvio Ciappi
ISBN: 978-1-4200-8693-5

Project Management Recipes for Success
Guy L. De Furia
ISBN: 9781420078244

Requirements Engineering for Software and Systems
Phillip A. Laplante
ISBN: 978-1-4200-6467-4

Security in an IPv6 Environment
Jake Kouns and Daniel Minoli
ISBN: 978-1-4200-9229-5

Security Software Development: Assessing and Managing Security Risks
Douglas Ashbaugh
ISBN: 978-1-4200-6380-6

Software Testing and Continuous Quality Improvement, Third Edition
William Lewis
ISBN: 978-1-4200-8073-5

VMware Certified Professional Test Prep
John Ilgenfritz and Merle Ilgenfritz
ISBN: 9781420065992

AUERBACH PUBLICATIONS

www.auerbach-publications.com
To Order Call: 1-800-272-7737 • Fax: 1-800-374-3401
E-mail: orders@crcpress.com

Solar Cell Technology and Applications

A.R. Jha, Ph.D.

CRC Press
Taylor & Francis Group
Boca Raton   London   New York

CRC Press is an imprint of the
Taylor & Francis Group, an **informa** business

AN AUERBACH BOOK

CRC Press
Taylor & Francis Group
6000 Broken Sound Parkway NW, Suite 300
Boca Raton, FL 33487-2742

First issued in paperback 2019

© 2010 by Taylor & Francis Group, LLC
CRC Press is an imprint of Taylor & Francis Group, an Informa business

No claim to original U.S. Government works

ISBN-13: 978-1-4200-8177-0 (hbk)
ISBN-13: 978-0-367-38509-5 (pbk)

Library of Congress Cataloging-in-Publication Data

Jha, A. R.
    Solar cell technology and applications / A.R. Jha.
        p. cm.
    Includes bibliographical references and index.
    ISBN 978-1-4200-8177-0 (alk. paper)
    1. Solar cells. 2. Photovoltaic power systems. 3. Solar batteries. I. Title.

TK2960.J54 2010
621.31'244--dc22   2009020578

Visit the Taylor & Francis Web site at
http://www.taylorandfrancis.com

and the CRC Press Web site at
http://www.crcpress.com

This book is dedicated to my beloved parents who always encouraged me to pursue advanced research and development activities in the fields of science and technology for the benefit of mankind.

Contents

Foreword

Fossil fuels such as coal, oil, and gas have been the primary means of energy generation for many centuries. However, thanks to global warming caused by burning fossil fuels, compounded by greater energy demand due to improved living standards and geopolitical tensions over oil resources and nuclear energy, we are facing a threat to the planet's well-being not seen since the last ice age. There is a new interest in renewable energy sources such as those derived from wind power, hydropower, or conversion of fast-growing crops into ethanol. These are more universally available and they can potentially help with global warming by reducing the carbon footprint. Then, there is a third kind of energy source based on photovoltaic conversion of sunlight into electricity by certain widely available semiconductors, which is arguably the cleanest, most ubiquitous, and potentially the most reliable alternative.

Dr. Jha's *Solar Cell Technology and Applications* addresses this important topic. The solar cell concept is a simple reverse biased p-n junction which converts absorbed light into electron-hole pairs and then into a small dc voltage. The cells may be stacked to charge a 12V automobile battery or to feed a power grid via a DC/AC inverter. Fortunately, among all the semiconductors suitable for this purpose, including newly discovered organic semiconductors, compound semiconductors such as CdTe, $CuInGaSe_2$, and GaAs, silicon is one of the most abundant materials in the earth's crust.

The book is divided into eight chapters which cover the whole gamut of solar cell technologies and applications:

- Chronology of scientific and technological developments since the invention of the solar cell at Bell Laboratories in 1954
- Design principles, equations, and models for bulk as well as thin-film cells
- Technology needs for residential, commercial, and utility power segments
- Certain universal techniques for enhancing solar conversion efficiencies
- Roadmap for the next two generations of designs and materials
- Solar panel design and technology for extraterrestrial applications

- Systems engineering for standalone and grid connected installations
- Relative economics of photovoltaic energy technologies over all the rest

Dr. Jha has published extensively on a wide variety of technical topics. This book continues to provide a testimonial to his unique ability to capture the essence of complex technologies in their entirety. Like the other books he has written, this volume also is closely grounded in real-life applications. He is addressing a problem that will only grow in importance over the coming years. It should help train a new generation of technical leaders well versed in the tradeoffs associated with multi-GW energy ecosystems. This book will be found most beneficial, particularly to students who wish to expand their knowledge on the subject concerned, project engineers, solar cell designers, solar power systems installers for residential and commercial applications, and space radiation-hardened solar power modules for space system applications.

Ashok K. Sinha, Ph.D.
Retired Sr. VP, Applied Materials, Inc.
Founder, SunPreme Inc.

Preface

Great interest in renewable energy sources and significant increases in the cost of foreign oil have compelled various countries to search for low cost energy sources and technologies, such as solar cells, wind turbines, tidal wave turbines, biofuel sources, geothermal technology, and nuclear reactors, to achieve lower cost for generation of electricity. This book comes at a time when the future and well-being of Western industrial nations in the twenty-first century's global economy depends on the quality and depth of the technological innovations they can commercialize at a rapid pace. Rapid development of low cost energy sources such as solar energy, wind energy, and tidal wave energy is not only urgent to reduce the cost per watt of the electricity, but to eliminate the dependency on oil-producing countries, some of which are hostile toward Western countries, particularly, the United States and European countries. It is important to mention that ample and free solar energy is available to countries located between the equator and Arctic Circle regions. Studies performed on solar intensities in summer and winter seasons indicate that a minimum NTP solar photon radiation intensity of 100 mW/cm^2 or 92.9 W/ft^2 is available during the sunshine periods, which could be converted into electrical energy with reasonably lower cost. Rapid design and development activities must be undertaken in the field of low cost and efficient solar cell devices, micro-hydro turbines, compact and light weight wind turbines, low cost tidal wave turbines, and solar concentrators. Design of low-cost, high-efficiency solar cell materials and light-weight, integrated solar panels must be the first priority to provide immediate relief to home owners and commercial shopping centers, which desperately need the solar power installations to reduce their electricity bills as well as to eliminate the dependency on foreign oil. Furthermore, energy experts predict that wholesale electricity prices will rise 35 to 65 percent by the year 2015. Under these circumstances, alternate electrical energy sources must be explored.

Solar cell technology has potential applications in satellite communication systems, military surveillance and reconnaissance space sensors, land-base defense installations, schools, residential homes, shopping markets, and large commercial buildings. Some Wal-Mart and K-Mart discount stores and Google already have operational solar power modules on their roofs. The solar installations could meet their electrical needs for lights, fans, air conditioners, computers, fax machines, water coolers, microwave ranges, and other electrical accessories. This book summarizes important aspects of solar cell technology critical to the design and development of solar power systems to provide electrical energy to various residential, commercial, industrial, and defense installations. Important properties of semiconductor and compound materials best suited for the fabrication of solar cells are summarized, which will provide optimum performance, improved reliability, and long operating life over 25 to 30 years. Techniques for performance improvements in solar cells and panels are identified, with emphasis on cost, reliability, and longevity. This book presents a balanced mix of theory and practical applications. Mathematical expressions and their derivations highlighting the performance enhancement are provided for the benefit of students who intend to pursue higher studies in solar cell technology. This book is well organized and covers critical design aspects representing cutting-edge solar cell technology. The book is written in language most beneficial to undergraduate and graduate students, who are willing to expand their horizons in the field of solar cell technology. Integration of organic dye technology in the development of solar devices is identified to achieve lowest electrical energy generation cost, improved reliability, and minimum fabrication efforts.

This book has been written specially for engineers, research scientists, professors, project managers, educators, and program managers deeply engaged in the design, development and research of solar systems for various applications. The book will be found most useful to those who wish to broaden their knowledge of renewable alternate energy sources. The author has made every attempt to provide well-organized material using conventional nomenclature, a consistent set of symbols, and identical units for rapid comprehension by readers with little knowledge in the field concerned. The latest performance parameters and experimental data on solar cell devices and solar modules are provided in this book; they are taken from various references with due credits to authors and sources. The references provided include significant contributing sources. This book is comprised of eight chapters, each dedicated to a specific topic and residential and commercial application.

The first chapter describes the chronological scientific developments and technological advances in the field of solar cell technology over the period from 1954 to the present. Potential silicon, III-V semiconductor compounds and organic materials best suited for the development of solar or photovoltaic (PV) cells are identified, with emphasis on cost, reliability, and conversion efficiency. Various reasons such as rising oil prices, terrorist attacks on oil installations, severe greenhouse effects, adverse political environments, and high transportation cost have compelled energy planners to look for alternate energy sources such as solar cell technology, which offers clean, environmentally friendly solar energy. Large amounts of carbon dioxide emissions are released only when electrical energy is generated using oil, gas, and wood, but no greenhouse gases are released when PV cells are used to generate electricity.

The second chapter focuses on the design equations involving critical performance specifications and design parameters. Materials best suited for second-generation solar cells, which will ultimately replace the traditional semiconductor materials such as silicon and gallium arsenide currently used in the first generation of solar devices, will be identified. Emphasis will be placed on thin-film technology most ideal for the next-generation solar cells to achieve low-cost, high-efficiency PV cells. Design improvements for gallium arsenide thin-film, monocrystalline silicon, and multicrystalline silicon solar cells will be identified, with emphasis on efficiency, power output, and reliability. It is important to mention that the spectral response of the solar device is dependent on the depth of the p-n junction, surface conditions, cell junction area, the wavelength of the solar incident light, and the absorption coefficient of the fabrication material. The adverse effects of surface roughness are discussed. A theoretical model is presented, which describes the mechanisms involved in determining the shape of the spectral response curves.

The third chapter describes the classifications of solar cell devices based on performance, design complexity, and manufacturing cost in dollars per kilowatt-hour. The studies performed by the author will indicate that for residential and commercial solar power system applications, it is essential to have both the low manufacturing cost and high conversion efficiency to meet the cost-effective criterion for deployment of solar cell technology. In case of military and space solar system applications, minimum weight, compact size, and ultra-high efficiency are the principal design requirements for the solar cells. The studies will further indicate that V-groove multijunction silicon (VGMJ) solar cell technology offers optimum design flexibility, lower fabrication cost, and high efficiency best suited for both residential and commercial solar energy systems. Critical fabrication processes such as plasma etching, laser ablation, passivation layer thickness, mechanical abrasion, and optimum patterns for rear contacts on the dielectric layer will be described, with emphasis on cost and reliability. Design techniques for lower spreading resistance and minimum contact resistance will be outlined. Design requirements and performance capabilities of amorphous silicon (a-Si), gallium arsenide (GaAs), copper-indium selenide (CIS), copper-indium gallium selenide, cadmium telluride (CdTe), Scotty-barrier solar, VGMJ, and multi-quantum-well (MQW) solar cells will be briefly summarized.

The fourth chapter focuses on potential techniques capable of enhancing the conversion efficiencies of the solar cells regardless of the device types and material used in the fabrication of the solar cell devices. Critical design aspects such as contact configurations and materials, front cover surface with optimum performance, antireflection coatings, oxidation process requirements for silicon surfaces, light trapping techniques, doped region thickness for lower recombination losses, grain size of the cell material, and absorption constant for improved spectral response. In addition, optimum surface layer thickness for high internal collection efficiency and fundamental collection efficiency will be specified. Benefits of bifacial modules, hemispherical mirrors and tracking mechanisms for achieving high concentration ratios, nanotechnology materials, including nanowires and nanocrystals, and all-dielectric microconcentrators are summarized, with emphasis on the improvement of spectral efficiency and conversion efficiency of the solar cell.

The fifth chapter defines the performance requirements for the second-generation and third-generation solar cells deploying advanced semiconductor compound materials and exotic structural configurations capable of yielding high conversion efficiencies. New design concepts and advanced cell materials will be investigated to achieve maximum conversion efficiency and reliability under severe operating environments such as space-radiation environments. Adverse effects of space radiation on solar cell performance will be identified. Note the conversion efficiency of the solar device can be derived in two ways, namely, the thermodynamic and the balanced principal. Preliminary studies performed by the author on solar cell efficiency parameters indicate that the practical efficiency of the solar cell will be far less than the theoretical efficiency limit imposed by the boundary conditions and operating environments. It is important to mention that the solar energy spectrum is very broad, ranging from the ultraviolet (UV) region to near-infrared (near-IR) region, whereas a semiconductor solar cell material such as silicon or gallium arsenide can only covert the photons with the energy of the band gap with optimum efficiency. The operational analysis will be performed, which will reveal that photons with lower energy are not absorbed and those higher energy levels are reduced to gap energy by thermalization of the photogenerated carriers.

The sixth chapter will describe the performance and design requirements of solar cells and arrays best suited for space applications. Critical design configurations and performance requirements for the solar cells and modules capable of powering the space-based surveillance and reconnaissance electro-optical sensors and communication satellites are described, with emphasis on reliability and uninterrupted system performance. Communication satellites over the last three decades have used silicon p-n junction solar cells for most of the space missions conducted by NASA, the Department of Defense and COMSAT Corporation. However, since early 2000, multijunction solar cells incorporating three or four semiconductor layers capable of absorbing energy levels in various spectral regions are getting the most attention for space applications. Design configurations for multijunction solar cells with high conversion efficiencies and output power levels in space environments are described, with emphasis on platform stabilization requirements. Critical topics such as solar array design and installation requirements, stabilization concepts, impact on solar energy system performance due to orbit fluctuations and space radiation, and reasons for reduction in electrical power level will be discussed in great detail.

The seventh chapter describes the critical system elements, installation requirements, and performance capabilities and limitations of stand-alone PV and grid-connected PV systems with emphasis on installation cost and complexity. Minimum electrical load requirement, geographical location, and average sunlight available per year are the principal considerations for any PV-based power system and they will be discussed for both systems in great detail, with emphasis on cost, design complexity, and reliability. It is important to mention that a standby battery package is required for a stand-alone solar power system, if electrical power is desired continuously for 24 hours. No standby battery package is required for a grid-connected PV-based power system. Benefits and disadvantages of stand-alone and grid-connected PV-based power systems are summarized. A stand-alone PV-based solar system is best suited for hot water applications and for heating homes or business offices with minimum cost and no generation of greenhouse gases. A computer program is identified that selects the solar array size for a given electrical load requirement based on continuous availability of solar energy, standby battery charge and discharge rates, and tilt angle requirement for uniform output power over the entire year.

The principal objective of chapter eight is to describe the performance capabilities and limitations as well as economic benefits of potential alternate energy sources, with particular emphasis on electricity generation cost per kilowatt-hour, greenhouse effects, system longevity, and overall system reliability. Installation requirements, capital investment, design complexity, and operational safety aspects will be identified for various alternate energy sources. Solar energy generation cost per kilowatt-hour will be compared with energy generation costs using other fuels such as coal, diesel oil, natural gas, wind turbines, hydroelectric turbines, wind turbines, tidal wave turbines, biofuel, geothermal technology, and nuclear fuel. Critical issues are addressed, such as energy conversion, fuel storage requirement, maintenance requirement, and continuous monitoring of critical performance parameters for safe system operation. Maintenance cycles and shut-down requirements especially for coal-fired steam turbines, hydroelectric turbines, nuclear power reactors, and tidal wave turbines are defined. Performance capabilities for various alternate energy sources are summarized, with emphasis on maintenance requirements, reliability, electricity generation cost, and installation cost. The amount of carbon dioxide generated by burning various fuels is summarized.

I wish to thank Dr. Ashok Sinha for his critical review of some of the chapters by accommodating last minute additions and changes to retain the consistency in the text. His suggestions have helped me to prepare the manuscript with remarkable coherency. Last, but not least, I wish to thank my wife Urmila Jha, who has been very patient and supportive throughout the preparation of this book, as well as my daughters Sarita and Vineeta, my son-in-law Anu, and my son Lt. Sanjay Jha, who inspired me to complete the book on time under a tight time schedule.

Chapter 1

Chronological History and Scientific Advancements in the Development of Solar Cell Technology

1.1 Introduction

This chapter describes the chronological history and scientific advancements in research and development activities pertaining to solar cell technology from 1954 to the present. Material scientists have identified potential III-V semiconductor compounds, nanotechnology and organic materials that can be used in the design and development of solar cells, also known as photovoltaic (PV) cells. It is important to point out that PV cells are not only environmentally friendly, but they offer clean, efficient, reliable, and uninterrupted sources of electrical energy. Why is the need for solar technology now so important? Rising oil prices, terrorist attacks on oil installations, high energy costs, adverse political environments, severe weather conditions, and global greenhouse effects have compelled energy planners to look for alternative sources to reduce reliance on fossil fuels such as coal, oil, and gas, and to switch to other clean forms of energy needed to protect our planet [1]. Note that large amounts of carbon dioxide emissions are released when electrical power is generated using coal, gas, and wood. Furthermore, coal extracted from remote coal mines must be transported to the coal-fired power plant locations. This involves huge transportation costs and delay in coal delivery under adverse climatic conditions.

Nuclear power plants are costly and take a long time to build. Recently a plan to construct two coal-fired plants in Utah was dropped, because of high costs and objections to global warming. A recent study by the industry-based Electric Power Research Institute projects that coal power generation will cost more than nuclear power generation or a natural gas generating power plant by 2025, even if carbon dioxide emissions are reduced to the greatest extent experts envision. Another industry analysis expert predicts that wholesale electricity prices will rise 35 to 65 percent by the year 2015, if the U.S. congress introduces a bill for a strict ban on greenhouse effects or carbon-based pollutants. Construction and installation of nuclear power plant involve a capital investment of more than $3 billion and takes a minimum time of two to three years. In addition, the disposal and storage of radioactive waste presents a serious problem. Electrical power generating plants can also use natural gas, but natural gas is volatile in both supply and price.

For the reasons cited above, energy scientists across the political spectrum have heightened interest in alternative energy sources such as wind turbines, hydroturbines, and solar cells. It is equally important to mention that the photovoltaic (PV) technology offers the most direct method to convert solar energy into electrical energy without carbon dioxide emissions or greenhouse effects. Solar energy is based on the photovoltaic effect, which was first observed in 1839. A PV device incorporates a p-n junction in a semiconductor material across which a voltage is developed from the solar radiation. The voltage generated across the junction is dependent on the properties of p- and n-materials and the diffusion constant. A one-dimensional theoretical model of a PV cell is shown in Figure 1.1. Sunlight absorption occurs in the semiconductor medium. Silicon is a weakly absorbing semiconductor material, thereby yielding the lowest collection efficiency. The semiconductor material used in the development of a PV cell must absorb a large portion of the solar spectrum closest to the surface to achieve high collection efficiency.

1.1.1 Chronological History of Developmental and Photovoltaic Power Generation Schemes Worldwide

Solar cells were first deployed to provide electrical power for space vehicles and satellite communication systems in the late 1950s, because these devices need no maintenance over long periods (5 to 10 years) and offer maximum reliability with no compromise in conversion efficiencies. Silicon solar cells were used to supply electrical power in the Vanguard satellite put into orbit in 1958. Thereafter, such

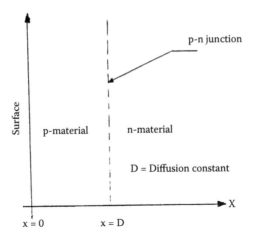

Figure 1.1   One-dimensional geometry for the theoretical model of a PV cell.

cells were frequently deployed in terrestrial satellite; nevertheless, space applications remained the principal market for more than two decades. Leading energy scientists recognized in 1973 that photovoltaic is a viable candidate for future nonfossil energy supply. Research activities on PV cells indicate that the development cost and system component cost have to be reduced by a factor of 1000, if solar cell technology is to be deployed for commercial and domestic applications. At present the price for grid-connected power systems currently operating in the United States, Japan, and European countries has been reduced by a factor of 100 approximately.

A 2003 survey of market shares of new technologies being used in the design and development of solar cells reveal that about 55 percent was accounted for by polycrystalline silicon technology, as shown in Figure 1.2, 30 percent by single crystal silicon technology, 5.6 percent by amorphous silicon (a-Si) technology, 6 percent by a-Si-on-CZ slice technology, and 3.5 percent by thin-ribbon technology. Scientific research studies on solar cells using thin films of cadmium telluride (CdTe) indicate that such devices suffer from high fabrication costs and low

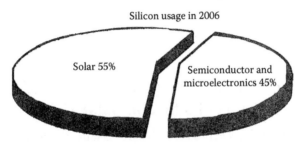

Figure 1.2   Solar cell applications now account for more than 55 percent of the silicon wafers used worldwide.

efficiency. Thin films of indium-tin (InSb) oxide are considered most attractive in fabrication of PV solar cells, but they also suffer from high cost, which can be significantly reduced using Q-switched, solid-state lasers operating at 1064 microns and using narrow pulses at high refreshing rates.

1.1.2 Why Solar Energy?

The demand for energy has always been the primary driving force in the development of industrial capability. The invention of the steam engine sparked the industrial revolution and the consequent evolution of an energy economy based on wood and coal. Since then the continuous growth of the energy economy has focused on various sources of energy, such as nuclear, wind, water, oil, and gas. Nuclear energy is very expensive and poses radiation hazards and nuclear waste problems. Electrical energy sources using coal, wood, gas, and oil generate large amounts of pollution or carbon dioxide emissions, thereby posing health risks. All these electrical energy power sources require large capital investments and scheduled maintenance. In case of coal-fired power plants, high capital investment, coal transportation cost, and delivery delay under adverse climatic conditions could pose serious problems. On the other hand, a solar energy source provides pollution-free, self-contained, reliable, quiet, long-term, maintenance-free, and year-round continuous and unlimited operation at moderate costs. Despite all these benefits of solar cells and nearly 55 years after their invention, PV solar cells are generating only 0.04 percent of the world's on-grid electricity due to the high cost of solar cells, which is beyond the reach of the common consumer. Based on the 2007 statistical review of world energy consumption, 30 percent of the electrical energy is generated from coal, 16 percent from natural gas, 15 percent from water generators, 9 percent from oil, 4 percent from nuclear reactors, and only 1 percent from solar cells. In the United States, solar energy of all kinds fulfills less than 0.1 percent of the electrical demand. All industrial and Western countries such as the United States, Germany, Japan, Brazil, Italy, Spain, and other European countries are turning to electrical power generation from solar cells, because of the high capital investments, radiation, and carbon dioxide emissions associated with coal-based, nuclear-based, gas-based, and oil-based power plants.

1.2 Identification of Critical Parameters and Design Aspects of a Silicon Solar Cell

Silicon-based PV semiconductor solar cells have been used to demonstrate the most practical and reliable application of the photovoltaic effect. A simple, rugged semiconductor junction can be produced from single-crystal silicon. Low resistance contacts are added to tap the electrical energy produced when the cell is exposed to sunlight. Approximately a DC voltage of 0.45 volts is generated across each cell regardless of the dimensions for this particular cell architecture. The DC current and thus the power available are strictly dependent on the cell area exposed to the sun and the absorption capability of the silicon wafer, which is located between the two contacts, as illustrated in Figure 1.3. It is important to mention that the higher the absorption capability of the semiconductor material, the higher the PV voltage will be across the cell terminals. In case of a material with weak absorption capability like silicon, most carriers are generated near the surface. Based on the preliminary calculations, one can expect DC output power ranging from roughly 250 mW from a 57 mm cell to about 1000 mW or 1 kW/m² from a 100 mm cell under standard conditions. Standard conditions are defined by NASA as 100 mW/cm² solar intensity (I) with cell temperature of 28°C at sea level. Higher voltage is

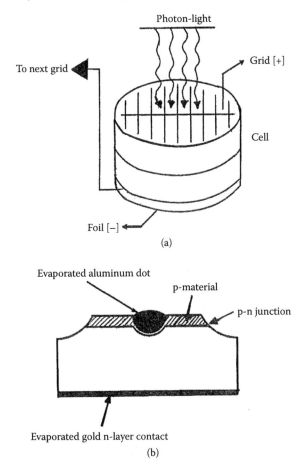

Figure 1.3 **Solar cell architectural details showing (a) schematic diagram and (b) cross-sectional view of the cell.**

possible by connecting the cells in series, while higher output power is possible by connecting the cells in parallel. Solar cells can be installed on glass-filled polyester substrate. The net cell output is the product of solar intensity ($I = 100$ mW/cm^2) and the conversion efficiency of the device, which is typically now about 16 percent for a silicon cell. A solar module may contain several cells connected in series and parallel using the most advanced interconnect and encapsulation techniques. The module design must offer the cost-effective approach to meet specific solar output power requirements with no compromise in reliability. These modules can be mounted on solar panels in series and parallel configuration to meet specific power generating capability. Solar cells must be packaged in a variety of modules to optimize the electrical performance of the solar power system best suited for a specific application. It is important to mention that basic solar cells are available with diameters of 57, 90, and 100 mm. Furthermore, half cells, quarter cells, or other configurations can be used where a specific application is warranted. Modules could use different packaging and encapsulating materials to achieve maximum economy and to meet different environmental requirements. For many applications, silicone is used to hermetically seal and to protect the cell integrity under harsh operating conditions. Polycarbonate or glass cases must be used to meet high impact resistance and maximum protection under severe mechanical conditions. All modules should be designed or constructed with minimum onsite maintenance and with self-cleaning capability for optimum shelf life. Other accessories such as voltage regulators, inverters, and mounting hardware are required to complete the solar generating system.

As mentioned earlier, a large number of solar cells and modules will be required to meet specific output power requirements. The base of the panel provides the mechanical integrity, while the glass cover offers maximum protection from environmental factors such dust, rain, wind, humidity, and suspended foreign particles in the atmosphere. This cover must provide the maximum transmission to the solar radiation, but with minimum reflection and minimum absorption losses. Typical panels commercially available have a length of 48 inches, width of 18 inches, and depth not exceeding 2 inches, and the assembled panel weighs less than 20 pounds. Computer analysis is necessary to achieve the most cost-effective panel design comprising the solar modules, also known as solar arrays, and the inverter. The analysis must specify the proper angle of tilt capable of optimizing both the location and solar system performance. Panel configuration, location, and installation requirements will be discussed in greater detail in a separate chapter.

1.3 Applications of Solar Power Systems

Since PV energy systems provide a fuel-free, pollution-free, and uninterrupted source of electricity, solar energy systems are suited for many applications, such as landing obstruction lights for airports, water pumping for irrigation, power sources for homes and commercial buildings, perimeter alarm transmitters, electronic border fences, intrusion alarms for security, highway signs, portable backpack radios, remotely located, unmanned electronic surveillance systems, educational TV broadcasting, railroads, radio relay stations, navigation aid sensors, ocean-based earthquake warning systems, emergency alarm transmitters, communication satellites, and space-based missile surveillance and reconnaissance systems. The most popular applications of the solar energy system will be discussed briefly.

1.3.1 Solar Power Sources for Homes and Commercial Buildings

Currently solar electricity generating systems for homes and commercial buildings are getting the highest attention. Recently, a San Jose office building owner David Kaneda [2] revealed the existence of an office building that consumes zero electrical power, generates no carbon dioxide emissions or greenhouse effects, and requires no fossil fuels for heating or air conditioning. The San Jose builder and his Santa Clara-based firm have remodeled the electrical and lighting systems incorporating the latest solar cell technology. He has installed solar panels on the building roof along with skylights in between the panels to illuminate the building with natural sunlight. Recently, the Santa Clara builder embarked on renovating older buildings with solar panels, with the goal of creating environmentally friendly residential and office buildings. It is important to mention that the skylights are sometimes supplemented with efficient fluorescent lights. Mr. Kaneda has disconnected the natural gas pipes for heating the building and recommended an alternate heating scheme, namely, solar electrical energy, by installing enough photoelectric panels to meet the entire electrical load of 30 kW, approximately. This 30-kW solar electrical energy will meet all the lighting, heating, and moderate cooling requirements of the building during the day. Three critical areas need to be addressed to cut down electrical energy use: lighting, heating, cooling, and accessories load such as the computers, printers, microwave ovens, refrigerators, and other things you plug into the wall.

Some building architectural aspects must be given serious considerations. The concrete parameter of the building must provide provisions for installation of windows and skylight necessary to cut down on the energy need. Special glasses must be used for windows to allow visible light through the glass but block the infrared and ultraviolet light in order to keep the office cool. An overhang on the south side will shade the window from direct sun on the east side, while an electro-chromic glass controlled by a sensor will darken the windows when sun hits them directly and will make them transparent the rest of the day. High ceilings will provide diffused light in the office and in areas where the skylight illumination is too strong, various diffusers can be deployed, if needed. Low-energy fluorescent bulbs must be used in conjunction with switching circuits and low-cost dimmers to save electricity.

Designers of energy-efficient buildings are recommending installation of geo-thermal heating pumps for cooling and heating the building. The geothermal heat pump takes advantage of the fact that at some depth below the earth surface, the ground temperature remains constant at 10°C or 42°F. This depth varies from place to place, but in Northern California, this depth is about 6 feet below the ground level. When the water flows into the building through the water pipes installed below this depth, it goes through a heat exchanger that collects the heat from the ground in winter and pulls the heat out of the building in summer. This way, the geothermal heat pumps provide heating and cooling in the building with mini-mum electricity. Thus, geothermal heat pumps can be operated with minimum solar generated electrical energy with no greenhouse effects. Electrical energy can be saved through several methods using LCD screens, light and motion sensors needed to turn off the lights when not needed, desk top computers, and energy-efficient appliances and office accessories.

1.3.1.1 Corporate Rooftops Using High Capacity Solar Energy Systems

Corporate rooftops are the latest frontiers in solar energy generation techniques [2] and have received the greatest attention from industry giants like Google, Applied Materials, Target, Kohl's, Wal-Marts, Tesco Supermarket in the United Kingdom, and corporations in Germany, Spain, Australia, Israel, Japan, and Brazil. In the United States alone, solar power generation capacity has increased from 3 MW in 2003 to 185 MW in 2007, approximately, and more than 70 percent of the solar installations were in California, because this state offers tax incentives. For example, Google's rooftop solar power plant involves 9222 polysilicon solar panels in the quasi-tilted horizontal plane facing west. The solar power system is capable of generating 9000 kWh of electricity before the sun fades into a flat orange ball and disappears into the Pacific Ocean. The solar modules blanket virtually all the free roof space on the eight buildings. Even the rows of carport roofs have more solar panels to produce additional amounts of electricity. Google claims all these panels generate 1.9 MW of electricity, enough to meet 30 percent of the buildings' peak electrical demand or to power one thousand California homes. Google's solar power installation is the largest in North America.

Corporate rooftops with solar energy generation capabilities are getting world-wide attention. Factories in Germany, Spain, Japan, and the Netherlands are involved in design and development of rooftop solar energy power sources. Markets for photovoltaic rooftop installations are increasing by 40 percent annually in the United States alone. Solar rooftop installation grew 100 percent in Spain in the year 2006. The German solar energy market was relatively flat in 2006, but Germany will install more PV generating systems in the very near future, according to a report by the Solar Energy Industries Association. California has become the second fastest-growing solar market in the world, which is being driven mainly by activity on corporate rooftop solar installations. In 2006, the commercial sector accounted for 66 percent of newly installed capacity in the United States, up from 14 percent in the year 2001, according to data released by the U.S. Department of Energy.

In March 2007, Applied Materials of Santa Clara, California announced a plan to install a 1.9-MW solar power generating system on the rooftops of its Sunnyvale, California complex. In the United Kingdom, Tesco, the British-based supermarket chain, intends to put up a 2-MW solar energy installation at an office complex. Wal-Mart, the world's largest retailer, intends to outshine all these companies with multipart plans to install solar power systems with generating capacity exceeding 5.6 MW on the roofs of 222 stores in California and Hawaii. Two other discount retail giants, Target and Kohl's, have immediate future plans to transform their roofs into tiny, independent solar energy power sources.

The corporate rooftop solar installations will realize significant increases in solar power generation, if improvements in solar panel efficiency and reduction in the solar panel cost are materialized. The current solar panel efficiency hovers around 15 percent (maximum) and the prices of solar modules are expected to decrease by about 5 percent a year. Based on these facts the industry energy experts predict that the solar energy systems will not be able to compete with electrical power sources offered by public utility companies until the year 2015 at least. However, with aggressive federal and state tax rebates and subsidies and passing strict environmental protection laws, the situation can change.

1.3.1.2 Solar Module and Panel Installation Requirements

Module performance requirements and solar panel cost play a key role in justifying the deployment of solar energy installations on corporate structures and commercial buildings. The power generating capability and conversion efficiency per solar module will determine the solar panel installation cost. In the case of the Google solar installation, each solar panel generates roughly 210 W using polycrystalline silicon cells, with each module having an efficiency better than 13 percent. Since the solar panels produce DC current, each system requires an inverter to convert the DC to AC current to comply with utility power supply requirements. Such converters have conversion efficiencies better than 96 percent. Energy experts estimate that the solar installation cost ranges between $3 and $5 per watt in California, and between $6 and $10 per watt elsewhere in the United States, after factoring in local and federal rebates and tax incentives for the solar system installation. According to the Northern California Solar Energy Association, the average cost of installing a large solar system in the Bay Area was $8.50 per watt before rebates and tax incentives, which can bring down the installation cost close to $2.80 per watt after various incentives. Based on these cost data, the Google solar system most likely will retrieve about $4.5 million from the state of California on its total investment of more than $ 13 millions. Federal tax breaks through the Energy Policy Act of 2005 will yield further savings to Google.

It is important to point out that aggressive research and development activities have so far failed to slash the solar installation per watt. Google was eager to undertake this solar power installation project, because Google owners have invested heavily in a start-up company called Nanosolar that specializes in thin-film solar cell technology capable of yielding solar cells with higher conversion efficiencies and lower fabrication costs. In addition, restrictions on carbon emissions and the volatility of the electricity prices have compelled the corporate giants to consider seriously the alternate and cheaper electrical energy sources.

The latest published reports reveal that solar power installations are getting financial incentives and assistance from solar service providers, which can run into millions of dollars. The solar providers are persuading prospective business customers to sign agreements that in effect turn those providers into miniaturized utilities companies. The office-supply company Staples was the first to pursue such a scheme in 2004 with Sun Edison, a prominent solar electricity service provider in Maryland for a 280-kW solar installation system on its two warehouses in California. This particular solar installation covered more than 10 percent of the electric loads for both facilities. Sun Edison installed the solar modules with no charge on the customer warehouses and took responsibility for the system maintenance. In this situation, the customer signed a long-term contract not exceeding 20 years, which locks the customer into buying back the electrical energy generated by those panels at a fixed rate. This rate is lower than retail prices charged by utility companies. Independent energy consultants reveal that 40 percent of the recent commercial solar installations have gone this way and this is likely to grow more popular as additional companies adapt solar energy installations. This scheme will be found most economical for companies that experience peak electrical load during business hours, typically 10:00 am to 5:00 pm. Because the electricity prices charged by utility companies are usually double, triple, or even quadruple, the solar electricity service provider scheme will be most cost-effective and the customers will be free from electricity price fluctuations and carbon emissions. It is important to mention that Power Light Company in Berkley, California, a subsidiary of SunPower, is one of the largest solar module producers and installers in the United States.

1.3.1.3 Impact of State and Federal Tax Rebates and Incentives

Energy planners predict that supplies of fossil fuels, natural gas, and crude oil will be significantly diminished within the next 15 to 20 years, while the demand for electricity has been doubling every 10 years, which will pose health hazards due to carbon emissions. Since the largest source of energy to the earth comes from the sun, which contributes 5000 times the total energy input from all other sources, it is most logical and desirable to use solar energy as much as possible. Furthermore, state and federal tax credits and incentives will accelerate the deployment of solar electricity provider schemes that are not only cost effective, but also are environmentally friendly, reliable, and independent of foreign restrictions and price fixing. Driven by these rebates in the past five years, several stores such as Target, Wal-Mart, and Home Depot are seriously considering 30-kW or more solar energy installation schemes. The California Solar Initiatives credit companies based on performance metrics that can amount to 25 percent of the cost of the solar electric system, including the cost of installation, solar panels, inverters, and mounting accessories. When the state and federal tax credits are taken into account, the businesses can recover more than 50 percent of the overall solar system cost. In 2007, California offered rebates based on solar system size rather than a per-watt basis and the rebate formula takes into account details of the physical placement of the solar panel, which allows reimbursements more generously subject to performance. The solar electric system performance is dependent on panel tilt angle and shading period as well as the altitude and azimuth coordinates commonly used to describe the sun's apparent position in the sky. The rebates are part of a tiered system designed to reduce the incentives over time and in the 2008 energy-efficiency requirements to these rebate dollars. In the United States only New Jersey has demonstrated great interest in pursuing solar installations based on photovoltaic technology and had the second-highest installed solar capacity of 18 MW in 2006. By comparison, Florida has a solar energy installation with moderate capacity of 170 kW. Arizona, New York, Colorado, Texas, and Massachusetts have indicated great interest in solar power systems.

1.3.1.4 Photovoltaic (PV) Installation Capacity Worldwide

It is interesting to note that the countries where solar panels have fared well are not always in the sunniest places in the world. Solar energy experts indicate that in 2005, three countries accounted for 90 percent of the 3075 MW installed photovoltaic capacity, which was upgraded to 4500 MW as of December 2007. Because of high electric, gas, and oil prices throughout the world, most of the countries in Europe and some in Asia are giving serious consideration to renewable energy programs such as solar electric energy. Particularly, the electricity prices are more than 20 cents per kWh or more per unit, which are two times higher than those in the United States. The disconnect between the sunshine and solar output is even more pronounced outside the United States, according to data published in 2006 by the International Energy Agency. Starting in the mid-1990s, both Japan and Germany began investing in renewable energy programs. As a result, today in cloudy Germany, the renewable energy industry has become the country's second largest source of new jobs after the automotive sector. Germany employs more than 200,000 engineers and scientists engaged in research and development activities with a major focus on solar energy programs. Worldwide photovoltaic installation capacities are summarized in Table 1.1.

Table 1.1   2005 Photovoltaic Solar Installation Capacity Worldwide

Country	Year Installed	Installed Capacity (MW)	% of World Total
United States	2004	480	12.9
Germany	1999	1,430	38.6
Japan	1994	1,425	38.4
Australia	2002	59	1.6
Spain	2003	59	1.6
Netherlands	2004	52	1.6

Note: The data presented here indicates the rapid growth of solar installation capacity, regardless of the geographical regions and climatic conditions. Furthermore, the data indicates that three countries, the United States, Germany, and Japan contributed 90 percent of the worldwide solar installation capacity in 2005.

1.3.1.5 Factors Impacting Solar Panel Installations

For corporations, the solar installation depends on factors other than rates and rebates. In the United States, rules vary from state to state regarding how to attach solar plants to the electric grid and how to compensate producers for electricity they export to the grid. There is a general rule of thumb that solar energy must generate less than 50 percent of the building's minimum electricity demand and this way the solar modules never come close to producing more electricity than the consumption needed for the building. Some utilities companies have been reluctant to open their grids to ever-larger quantities of the electricity that they cannot manage due to grid reliability and frequency or voltage disturbances. Some companies plan solar installation for 100 MW or up, but the slow response from the utility companies force them to scale down the project, which essentially restricts the growth of solar installation capacity. Electricity charges per kilowatt hour can block the expansion of solar installations. When the electricity prices are among the highest in the country, for example, 15 cents per kilowatt hour for the commercial sector, compared with a national average cost of, for example, 9 cents per kilowatt hour, companies will not go for the solar energy sources due to high cost of solar installation systems. Furthermore, project developers are not prepared to invest in solar installations until they are confident the utility company will agree to connect them hassle-free or without re-conditions. All these problems must be faced by the solar project developers before making a firm decision whether to go forward or not. Only a few states in the United States have adopted uniform interconnection standards for distributed solar power sources. Most of the utility companies have established their own interconnection requirements, which can be in conflict with those defined by individual states. In addition, some states have adopted the IEEE standards and guidelines defined by the National Renewal Energy distributed with emphasis on reliability and safety. Environmental conditions will greatly improve with rapid acceptance of photovoltaic installations. It is interesting to mention that Oregon set a policy in June 2007 that can amount to a 50 percent tax credit for solar installations and manufacturing to encourage the full use of statewide solar energy. Such moves can significantly bring down the prices of solar modules and installation costs. Solar energy experts predict that solar panel prices will see significant reduction as soon as the new manufacturing capacity comes online starting in 2008.

1.3.2 Photovoltaic Solar Energy Converters for Space Applications

Solar energy is the largest source of energy available to the earth, contributing five thousand times the total energy input from all other sources, and is reliable and environmentally friendly. Solar energy is a widely distributed source, which provides 1 kW of solar power per square meter of the earth's surface exposed to direct sunlight at noon. Photovoltaic solar arrays are best suited for communication, surveillance, and reconnaissance satellites, where a continuous, reliable, and environmentally friendly energy source is of critical importance.

Design aspects and requirements for solar cells, solar arrays and panel installations for satellites or space vehicles will be quite different from those used in earth-based solar energy systems. Solar cells and modules must have higher conversion efficiency, reliability, and mechanical integrity over extended periods ranging from 8 to 12 years under space radiation environments. Solar cells with n-p junctions are preferred over p-n junction cells because of higher space radiation resistance. Single-crystal solar cells are widely deployed in communication satellites and space vehicles. A solar array consists of a large number of cells, which are arranged in parallel- and series-strings to provide desired power and voltage. Blocking diodes are used to separate the strings to avoid reliability problems. The solar cells are bonded directly to the substrate, which consists of two face sheets of epoxy-fiberglass bonded to an aluminum honeycomb core. Proper adhesive materials are used to mount the solar cell modules to the substrates, which will be subjected to wide temperature variations when going into eclipse. The wide temperature excursions and requirements for long operating cell life required stringent design specifications for both the solar module and solar panel. Body-mounted solar array temperatures of spinning satellites are typically in the range of 32°F to 68°F except when the sun vector is parallel to the spin axis, as illustrated in Figure 1.4, in which case the temperature of the continuously illuminated solar panels reaches about 176°F in earth orbit. The panel temperatures will be slightly higher at low altitudes due to earth albedo (electromagnetic radiation).

The solar module and panel design and performance requirements will be very stringent for the next generation of satellites placed in synchronous equatorial orbits.

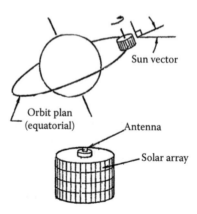

Figure 1.4   Schematic of a spinning satellite showing the locations of the antenna and solar array, as well as characteristics of the orbit.

The solar module and array design configurations will demand strict reliability and sustained performance under wide temperature variations. The trend toward lighter solar arrays and higher earth orbits will result in ever wider temperature excursions from sunlit to eclipse periods. These wide temperature excursions, long operating life requirement, and higher reliability will put great demands on the design and reliability of intercell and intermodule connections in the solar array. Molybdenum and Kovar are best suited for interconnections between the individual solar cells in the arrays, because they have expansion coefficients near that of silicon. These materials must be selected with regard to the temperature cycling. In the case of spinning arrays, which experience lesser temperature extremes, silver or copper interconnections can be used. In the latest satellite arrays, welding techniques are used instead of soldering to interconnect solar cells.

Space radiation damage to solar cells can be caused by charged-particle environments, including both trapped radiation and solar flare. The space radiation can affect the transmittance of the solar cover slides, cell-to-cover adhesive and the diffusion length of the minority carriers in the semiconductor, which can degrade the performance of the solar cell. The cover cell is the principal source for the protection of the solar cell. The effect of the radiation environment can be most serious on the solar cell power output capability. The radiation environment typically consists of electrons and protons with intensities and spectra varying widely as a function of orbit altitude, satellite inclination, and solar activity. The combined effects of these factors tend to degrade the electrical properties of the solar cells, which in turn will reduce the power output of the cells.

Outer space temperature effects can degrade the solar cell performance in terms of efficiency, power output, and open-circuit cell voltage. Solar cell voltage decreases with increasing temperature, which will lead to reduction in available power output from silicon cells roughly by 0.5 percent per degree centigrade. This means a solar array whose back surface can radiate to space will be relatively cooler and, therefore, will have a higher output power than an array whose back surface is blocked by the spacecraft or other equipment.

The solar array can be designed to almost any voltage at a given power level. However, there are concerns regarding reliability and power loss at a given voltage and power.

Power conditioning complexity can be relieved using the design of hybrid-arrays, which uses arrays with separate circuits for high and low voltages. Hybrid-array can reduce the weight of power conditioning system as much as 25 percent.

The power output from a solar array is optimum at the beginning of the life, which will degrade slowly over the subsequent years depending on the severity of the radiation damage, frequency of temperature excursions, orbit inclination, and altitude of the satellite. Communication satellite-based solar systems have power ratings ranging from 5 to 12 kW in the 1970s, 100 kW to 200 kW in the 1980s, and 300 kW and up in the 1990s and beyond. Solar power systems with higher power capacities are generally used by multichannel (voice and data transmission) communications satellites and reconnaissance satellites for specific military missions and are very complex and heavy. Communications satellites are spin-stabilized in synchronous equatorial orbits and their solar arrays are spinning cylinders, as shown in Figure 1.4. As the satellite power requirements increase the trend will be toward extendible arrays on three-axis stabilized spacecraft, as illustrated in Figure 1.5. In future satellite arrays, solar cells using thin films of GaAs or cadmium sulfide will be deployed, because they will yield high conversion efficiency, structural integrity, and enhanced space radiation resistance.

1.3.3 Radio Relay Stations

Microwave/ultra high frequency (UHF) communications satellites, central repeater stations atop mountains, in deserts, and remotely located pipelines are not easily accessible. Deployment of solar energy systems will allow complete freedom to such site locations and will eliminate the need for continuous fuel transport and mechanics to these remote sites, thereby avoiding costly maintenance trips and continuous presence of operating personnel. The latest market survey indicates that solar power generators are currently deployed to power communications equipment, environmental monitoring sensors, and control systems in the United States, Canada, Japan, Fiji, Italy, Australia, New Guinea, Europe, and other countries.

1.3.4 Navigation Aid Sensors

Solar electric power generators are widely used to service the airport landing lights, fog warning horns, and bells located on unmanned oil rigs, coastal buoys, and other marine installations in areas such Boston Harbor, the Gulf of Mexico, New Zealand, and offshore regions of Scotland and Venezuela. Note the reliability, ruggedness

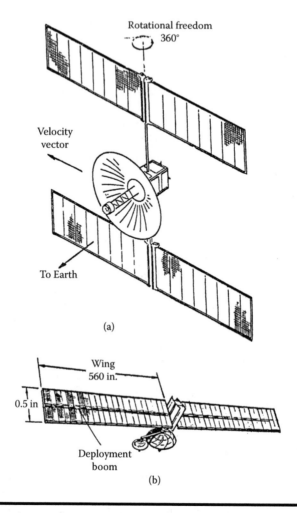

Figure 1.5 **Design configurations for TV broadcast satellites consisting of (a) a 12-kW solar array and (b) a 5 kW solar array.**

of solar power generators, and protection against corrosive environments must be principal design requirements for such use of solar power generating systems. These solar power generating systems will require no or very little maintenance, thereby yielding tremendous savings over the traditional battery-powered systems.

1.3.5 Railroad Communications Networks

Railroad communications networks, which provide the right-of-way capability to trains, are another important application. Solar generators are critically important to reduce hazard at unprotected railway crossings by providing reliable and continuous electrical energy for all lights, warning bells, and gates required to protect these crossings around the clock. Based on a 2006 survey, there are roughly 185,000 unprotected railway crossings in the United States alone. Track circuits, semaphore signal, and railroad safety devices that indicate the presence of a train on a certain section of track can be powered by solar power generators, which can replace the battery systems and conventional power sources. The reliability and economics of the solar power generators provide major benefits over traditional power sources such as diesel generators and thermoelectric generators. Railroad companies in the United States, Canada, Mexico, and other industrial countries depend on solar energy generators to power these critical operational systems.

1.3.6 Educational TV Programs

Currently solar energy systems are providing educational TV programs to school children in African villages and other third world countries. TV signals are available from both the urban broadcasting systems and satellite repeaters. Conventional remote receiving power systems such as batteries and diesel engine generators are too expensive, not cost effective, and require intermittent maintenance to ensure continuous supply of electricity. It is interesting to mention that educational TV stations are fully operational in Niger, Ivory Coast, and many other African countries. Solar energy system configurations for satellite-based TV broadcast arrays with 5- and 12-kW power generating capacity are shown in Figure 1.5.

1.3.7 Optimization of Solar Electric System for Specific Applications

Optimization of solar electric system can be achieved for a specific application by taking into account its maximum electric load requirement, geographical location, space available for mounting the solar panels, average sunlight available without shadow, and the environmental conditions in the vicinity of the solar electric system installation such as weather, pollution, seasonable factors, and surface reflectivity. A specially designed computer program is available to organize, integrate, and interpret critical solar system design parameters. This computer program can provide the most reliable and cost-effective solar electric system configuration involving the least expensive combination of solar modules incorporating the most efficient solar cells, inverters, and emergency storage batteries to ensure continuous, year-round operation for a certain electric power requirement and installation hardware.

Using the computer program, the solar electric system project engineer is able to select the smallest solar array size for continuous load and then calculate the electrical power output for every day of the year based on the availability of average sunlight per year, known as insolation. The power output can then be compared with the load demand per day and the difference is multiplied by the number of days when the solar array output is below the design load requirement. From these data, one can determine the number of days of storage required to ensure the continuous operation of the electrical load. After computation of the number of days of storage for a specific array size, the computer selects other closely rated array sizes and repeats this computer routine, until it determines the most economical solar electric system configuration. The computer program determines the normal battery discharge rate, maximum possible battery discharge rate, charging efficiency, and annual battery self-discharge. In addition, operational parameters such as taxes, the investment cost, and the projected battery life can be factored in whenever warranted.

The computer also determines the angle of tilt for the solar modules that yields the highest collection efficiency and the most uniform electrical output over the year. The tilt angle is calculated to smooth out the insolation versus time-of-the year curve and to achieve the maximum contribution to the least-cost determination of array size and the emergency battery output power requirement.

Some practical examples of solar power sources include the following. A flashing light in Boston Harbor, Massachusetts, has a daily electrical load requirement of 10 ampere-hours at 12 V. Preliminary calculations indicate that an array of 28 solar modules with three 100 ampere-hour rating from 12-V batteries will be enough to provide electrical power continuously at this location. The calculations further indicate that the same load in Marseilles Harbor, France, would require 28 solar modules with two batteries, the Tampa Harbor, Florida, would require 21 solar modules with one battery, and the Fairbanks Harbor, Alaska would require 35 solar modules with six batteries each rated at 12 V. It is quite evident from these three examples that solar module requirements vary from one geographical region to another based on climatic conditions and availability of continuous sunlight.

It is important to point out that optimization of solar power source is strictly dependent on the modular longevity, cell efficiency, material cost, fabrication complexity to incorporate solar cells into modules, and installation. There is no doubt that the solar cell technology benefited significantly from the high standard of silicon technology developed originally for transistors by Bell Lab scientists in 1954. The silicon technology dominated the market for over three decades. However, the research and development activities on group III-V compound semiconductor materials between 1980 and 2000 have identified alternate material sources for solar cells. The compound semiconductor materials include gallium arsenide (GaAs), indium-phosphide (InP), copper indium di-selenide (CIS), copper indium-gallium di-selenide (CIGS), and cadimium telluride (CdTe). Important properties of solar cell materials will be discussed in greater detail elsewhere. Materials with high absorption capabilities are best suited for solar cells and the critical properties of such materials will be identified later.

1.4 Fabrication Materials for Solar Cells and Panels

Brief studies performed by the author indicate that the 2006 solar cell market share includes 55 percent of polycrystal, 30 percent of single-crystal silicon, 6 percent of amorphous silicon (also known as a-Si), 5.8 percent of a-Si on Czochralski (Cz) single-crystal, and 3.5 percent of ribbon-silicon. Cadmium telluride was used first to fabricate solar cells in 1960, but its use was discontinued due to its extremely low absorption capability. The studies further indicate that high light absorption capability is the critical requirement for solar cell materials. High absorption capability is possible with thin films of group III-V compound semiconductor materials, such as GaAs, InP, CIS, CIGS, and CdTe. The use of CdTe material for fabrication of solar cells cannot be justified due to its poor absorption capability. Solar cells fabricated with thin films of CIS and CIGS have demonstrated conversion efficiencies better than 20 percent under laboratory conditions, which will be reduced to 18 percent or so. These solar cells are currently available in small quantities, but at higher procurement costs due to limited production. The studies further reveal that multijunction solar cells with V-groove configurations and using thin films of CIS and CIGS can boost the conversion efficiency better than 30 percent in the near future. Aggressive research and development activities on concentrated solar cells or PV cells can extend the efficiency close to 40 percent under laboratory controlled environments. Based on these efficiency projections, the roof-mounted multijunction solar cells will be able to generate electricity in the tens of kilowatt range with grid-connected power supply system with acceptable cost. Research and development activities have been pursued aggressively by Stanford University professors and other industry scientists on solar cells fabricated from light-weight, flexible, thin films of organic materials. However, reliable data on manufacturing yield and efficiency are not available at the present moment.

There are three basic types of solar cells that are widely used in solar generating systems. They include crystalline silicon cells with conversion efficiencies ranging from 15 to 22 percent, multijunction GaAs solar cells with efficiencies ranging from 26 to 29 percent, and various kinds of thin-film solar cells with conversion efficiencies in the range of 32 to 38 percent, approximately. Note the multijunction GaAs-based solar cells are fabricated using metal organic chemical vapor deposition techniques on monocrystal wafers with maximum theoretical conversion efficiency approaching 40 percent approximately. It is important to mention that these multijunction cells are very expensive to manufacture and, therefore, are limited to applications where cost is not an issue or where small areas are involved, such as satellites or solar concentrators.

1.4.1 Crystalline Silicon Solar Cells

Crystalline solar cells are the most widely deployed in many applications and are made from either mono- or multicrystalline silicon wafers with theoretical efficiencies ranging from 13 to 22 percent under laboratory environments. Solar cells using thin films of silicon have conversion efficiencies in the range of 8 to 17 percent and have lower production costs. Approximate relationship between the relative costs and efficiencies is shown in Figure 1.6 for thin films, silicon and gallium arsenide solar cells. Growth projections for the photovoltaic capacity for thin-film and silicon wafers is also shown in Figure 1.6. Note cost versus cell efficiency is a constantly changing picture as research scientists and innovators strive to increase the efficiency and to lower production costs of solar cell technology.

Silicon material in any form is best suited for the fabrication of solar or PV cells due to its minimum cost. Crystalline solar cells account for more than 94 percent of the photovoltaic market in the entire world. Solar cells can be fabricated from

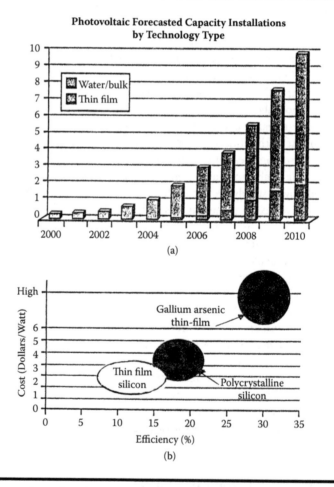

Figure 1.6 Solar cell technologies showing (a) installation capacities for various materials and (b) cell cost and efficiency for various cell materials. (Data from Reference 1.)

either mono- or multicrystalline silicon wafers, which have demonstrated conversion efficiencies in the 14 to 22 percent range. In 2006, solar cells surpassed the semiconductor industry in its use of silicon wafers, which account for some 55 percent of consumption worldwide, as shown in Figure 1.2, while only 45 percent of the silicon is used for the fabrication of microelectronics and solid-state devices.

The significant demand for silicon wafers for other applications has created a shortage in the supply and has pushed the price for silicon solar cells higher. This rise in silicon solar cells has motivated the PV industry to look seriously for alternate sources, such as GaAs thin-film technology and organic thin films, which use very little or no silicon. Thin-film solar cells generally consist of multiple layers of different materials coated onto low-cost glass or metal substrate. A typical solar cell structure is illustrated in Figure 1.7. The most common absorbing materials used include amorphous silicon (a-Si), cadmium telluride (CdTe), and copper indium gallium di-selenide (CIGS). Transparent conductive oxides such as zinc oxide or

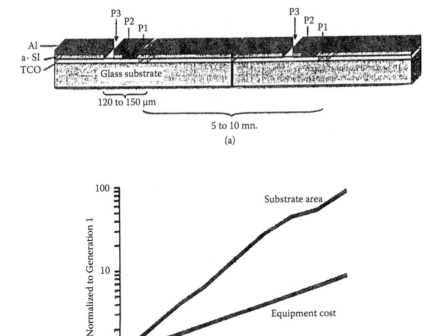

Figure 1.7 A thin-film solar cell. (a) Cross-section of a thin-film amorphous-silicon solar cell involving P1, P2, and P3 laser scribe patterns. (b) Manufacturing equipment costs as a function of substrate cost and solar panel size.

indium tin oxide and metals such as aluminum or molybdenum are used to form the outer electrodes of the solar cell. Solid-state lasers such as diode-pumped lasers operating at a wavelength of 1064 nm are used to scribe interconnect and isolation patterns at various stages during the manufacturing of a solar panel. Specific details on a typical a-Si solar cell can be seen through the schematic cross-section of the cell as shown in Figure 1.7.

1.4.2 Fabrication of a-Si Thin-Film Solar Cells Using Laser Scribing

Amorphous silicon-based thin-film solar cells can be installed on solar arrays with minimum cost and complexity. The first step in creating thin-film solar cells requires coating the front electrode onto the glass substrate, as shown in Figure 1.5. Indium tin (InSb) oxide coating is usually preferred, which is a transparent conductive material. Then the P1 electrode pattern is scribed using a Q-switched solid-state neodymium-doped vanadate operating at the fundamental wavelength of 1064 nm and using an average power between 12 and 15 W. It is necessary to scan the laser beam rapidly and to operate the laser at a high repletion rate at 100 kHz or more. Narrow laser pulse width of 15 to 30 ns is required to ensure that the laser peak power is above the material ablation threshold. Laser beam quality and pulse-to-pulse stability are necessary for a clean scribe and a reliable and repeatable process. Research scientists are investigating other wavelengths and higher repetition rates to meet faster throughput requirements.

After the indium tin oxide pattern, the solar panel or array goes back into the chemical vapor deposition machine, where it is coated with a thin layer of amorphous silicon semiconductor. This is then patterned with P2 scribes using a green 532-nm neodymium-doped vanadate laser from the back through the glass substrate. Again high repetition rates and short pulse widths of 15 to 30 nm are used to meet fast throughput requirements. In this case, the laser average power is less than 1 W. Note higher power lasers with average power from 4 to 6 W can be used to perform multiple laser scribes simultaneously, which will reduce significantly the manufacturing costs for solar arrays or solar panels with large surface areas. In other words, using this approach, the manufacturing cost per unit area is significantly reduced as the size of the panel increases.

Once the P2 scribes are completed, the solar panel is coated with aluminum, and then laser scribed with the final pattern, P3. This pattern is done from the back through the glass with a laser identical to the one used for the P2 process. Again, laser beam quality and laser pulse-to-pulse stability are critical for a reliable and good scribe and to avoid damage to the underlying layers.

1.4.3 Automated In-Line Processing for Thin-Film Solar Cells

The major advantage of the thin-film solar cell technology over crystalline silicon solar cell technology is that it is scalable and lends itself to automated in-line processing, which is essential for significant cost reduction of large solar panels. Large solar panels are best suited for commercial buildings. Processing companies that manufacture the in-line coating machines for the flat panel displays have recognized this opportunity and have already set up production lines in anticipation of large orders. The factory operation is exactly the same as in the display market and scaling up in the substrate area can be accomplished with minimum equipment cost. However, then manufacturing scales are done at different rates because of the large size of the solar panels.

It is important to mention that laser systems are of critical importance in in-line production systems. The laser plays a critical role in bringing to reality a technology that has shown great promise in the mass scale development of solar panels or arrays with minimum cost and complexity. Solid-state lasers operating at 1064 nm and employing short pulses at high repetition rates are not only cutting the production costs of solar arrays, but are also brightening the future of the solar industry.

1.4.4 Thin-Film Photovoltaic Market Growth

Solar industry innovators have predicted a bright future for the thin-film photovoltaic (TFPV) market. Scientists are predicting the market will reach $7.2 billion by the year 2015 [2], compared to just more than $1 billion today, according to a new market report from NanoMarkets LC, a solar industry firm in Glen Allen, Virginia. The strong TFPV market forecast is driven by the inherent advantages of the TFPV technology, such as low production cost, light weight, and ability to manufacture these films on low cost flexible substrates, which will lead to integrating solar power capabilities into walls, roofs, and windows with large surface areas. It is of critical importance to point out that TFPV cells have the ability to operate under low light conditions with reasonably good conversion efficiencies, whereas solar cells using crystalline silicon wafers are unable to operate under such conditions. The technical report generated by NanoMarkets [2] reveals that most solar panel suppliers are ramping up production capacity to support the growing demand for TFPV cells using nanotechnology. Companies such as First Solar, Fuji Electric, Nano-Solar, Uni-Solar, Terra-Solar, and Sanyo are building plants with more than 100 MW capacity to meet electrical power requirements. Installed solar power capacity (MW) as of 2005 for the United States, Japan, Germany, and Spain is shown in Figure 1.8.

The TFPV market looks extremely promising, because worldwide energy prices are rising very rapidly, photovoltaic cell prices are falling at a rapid rate and pollution is threatening human lives worldwide. Under these conditions, the PV market will become a big slice of the energy market and could ultimately account for more than 20 percent of the U.S. market's energy needs. TFPV market growth worldwide from 2007 to 2015 can be expected to grow by 20 percent or more from the installed solar power ratings shown in Figure 1.8. Because TFPV cells cost less than conventional PV or solar cells, TFPV technology-based cells will be most

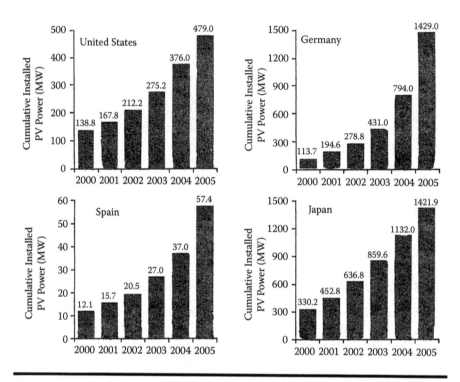

Figure 1.8 Cumulative solar installed capacity (MW) for various countries. (Data from Reference 6.)

likely to take off first on assembly lines. Just five years ago, TFPV demand was only 5 percent of the entire PV market. However, now it appears that the TFPV demand is expected to account for 35 percent of the photovoltaic market by 2015. Manufacturing costs for conventional PV cells are much higher than those for TFPV cells due to simple printing machines and material processes. It is expected that TFPV printing has the potential for lowering capital costs at least by 75 percent, reducing waste, improving quality control, and increasing throughput.

The fact that TFPV is very flexible and much lighter than the conventional solar cell makes it more easily applied to curved and nonplanar surfaces, thereby making solar panel installations on roofs and walls with minimum cost and complexity. Where many TFPV-based panels need to be installed, the roof will not require reinforcement, thereby realizing significant reduction in installation costs and enhancing the structural integrity of the roof. In addition, TFPV-based panels are equally suited for window surface installation, which makes the TFPV technology more cost effective as well as practical for commercial buildings irrespective of size.

The latest research and development activities on materials indicate that PV-based technology using organic materials offers hope for the future of solar cells, because they are more ecologically friendly than conventional PV technology.

Currently, the efficiencies of PV cells are not impressive; nevertheless, they are improving at faster rates. New architectures of organic PV cells promise the electrical performance could come close to or possibly even exceed those of their inorganic counterparts within 5 to 10 years. The latest market survey predicts that by 2015 shipments of organic PV cells could reach greater than 500 MW. Significant demand for encapsulation materials, transparent conductors and silicon ink can be expected, when demand for organic TFPV cells and organic/hybrid TFPV cells picks up in the near future.

Users of TFPV cells include large utility companies, commercial and industrial buildings, satellites, remote military installations, and emergency power systems to provide electricity under harsh environmental conditions. Several companies in the United States, Europe, and Japan are gearing up to meet solar power requirements using basic TFPV and hybrid technologies. Such companies include Antec Solar, Dow Corning, Global Photonic Energy, Shell Solar, Terra Solar, Uni-Solar, Wurtz Solar, Mitsubishi, International Solar Electric Technologies, First Solar, Solar Cells, and a host of other companies deeply involved in the manufacture of solar cells and solar panels.

1.5 Concentrated Solar Technology

Alternate concentrated techniques are being exploited to provide solar energy at minimum cost and complexity by boosting the conversion efficiencies to exceed 30 percent or more. Concentrated technology is necessary, because the theoretical conversion efficiency for single-junction silicon solar cells is limited to 16 percent, to 28 percent for normal gallium arsenide solar cells, 26 percent for reversed gallium arsenide solar cells, to 19 percent for copper indium selenide (CIS), 16 percent for thin-film cadmium telluride (CdTe) solar cells, 26 percent for V-groove silicon multijunction solar cells, and 45 percent for nanotechnology-based solar cells. When internal collection efficiency, fill factor, optical loss, contact loss, recombination loss, and reflection loss are taken into account as illustrated in Figure 1.9, these theoretical efficiencies will be reduced by a minimum of 3 to 5 percent depending on the sun concentration factor, cell junction temperature, and climatic conditions. But the net solar cell efficiencies may not be sufficient to generate electrical power to meet specific electrical load requirements. Note the solar energy source may not be able to provide electrical power at peak demand times during the night. Solar concentrated photovoltaic (CPV) technology provides both the cost potential and toxic-free operation [3]. In brief, high sun concentration factor, both the solar cell efficiency and power output increase. Cell longevity under high concentration is a matter of serious concern. Extensive laboratory test data on the cell's longevity must be collected as a function of sun concentration factor. Cell longevity is the most critical issue for the entire solar industry and offers both the most potential

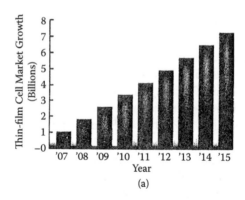

(a)

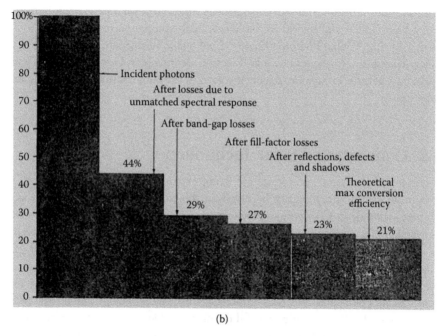

(b)

Figure 1.9 (a) Market growth for TFPV cells. (b) Impact of various losses on cell efficiency.

and catastrophic risk in capturing the power of the sun for electricity generation without carbon emissions and other harmful pollutants.

CPV technology embodies many different disciplines such as optical physics, material science, engineering, and manufacturing. On the optics front, a SolFocus company in Sunnyvale, California, has made significant progress in taking theoretical nonimaging and other optical design limits to an economical manufacturing state. This has required a significant investment in the glass business that will pay dividends as one realizes the tremendous cost-performance benefits from these approaches. Pressing glass to optical quality and precision while maintaining minimum cost under high volumes has been quite challenging.

Material science for wide spectrum antireflective coatings and diffraction gratings is critical. The multijunction solar cells exploit the entire solar spectrum but the antireflective coatings for this range are just coming to fruition. Thermal management is another area that requires improvement. One can take advantage of state-of-the art semiconductor fabrication processes to optimize solar cell performance and reliability with minimum cost. One should pull technical innovations from the aerospace, automotive, electronics, and semiconductor industries to improve the CPV technology best suited for solar energy sources.

SolFocus established several installation sites in 2007 to demonstrate the feasibility of the CPV technology concept. Although these sites are generating electricity for the utility grid, the primary objective of site installation was to provide performance data and reliability test information on the CPV cells. These sites will benefit in determining to drive costs out of the balance of system items including the installation, which represents a large share of the total solar power plant using CPV technology.

One pilot solar plant located in Sunnyvale is not only assisting with the initial production phase, but also refining the CPV assembly and test procedures. SolFocus is currently focusing on the installation of CPV systems in foreign lands. The company is moving tooling to India for production of CPV solar systems. In addition, SolFocus installed a 200-kW CPV power plant in Spain in September 2007 and a 300-kW plant there in early 2008 [3]. SolFocus recently acquired a Spanish-based sun-tracker company to manufacture mass scale trackers with minimum cost, which will further accelerate the installation of CPV-based solar power plants throughout the world at affordable prices.

1.5.1 Collaboration Key to Successful Entrepreneurship

Collaboration with foreign companies will lead to successful entrepreneurship in the CPV technology area. Moser Baer Photovoltaic (MBPV) is a unique partner, because MBPV is a reliable investor, an experienced manufacturing partner, and exclusive distributor for the SAARC (South Asian Association of Regional Corporation) region consisting of eight countries centered around India. MBPV shaved significant cost in controlling the thickness of silver sputtered on discs to within a micron. In addition, MBPV has the necessary know-how to meet the demanding reliability and quality standards with minimum labor costs.

In addition, SolFocus has collaborated with national and international research institutions such as the University of California at Merced, the National Renewal Energy Laboratory, Palo Alto Research Center, United Technologies Research Center, Ben Gurion University, and Madrid Polytechnic University. The company intends to enter further agreements with other national and international research partners and large energy companies actively engaged in the CPV technology.

SolFocus's principal objective is to continue to pursue the mission of delivering solar electricity at cost-parity or less than the energy generated from fossil fuels, which are the largest contributors to carbon emissions.

1.5.2 Low-Cost Concentrator Technique to Intensify the Sunlight

The CPV technique discussed earlier involves individual concentrating elements for each solar cell installed on the panel, which will lead to higher fabrication and panel installation costs. The optical concentrator technique involves a unique solar panel design configuration that magnifies the sunlight and the panel design has been optimized by the Israel National Center for Solar Energy. The Israeli research scientists claim that the new solar power development scheme would significantly reduce the high cost of solar power generation using solar panels. Each solar panel has a simple, low-cost reflector that is made up of several optical mirrors to intensify the sunlight a thousand times. The intensified light is so strong that it can burn a person. This new concentrator technique eliminates the need for more solar cell area. A 10-cm diameter sunlight receiver could intensify light and produce electrical energy that is equivalent to the energy level produced in a conventional receiver with a diameter of 10 m. Scientists predict that mass scale production of such panels would reduce the solar power source. This technique will not only relieve the dependence on fossil fuels, but also will help in significant reduction in carbon emissions.

Sophisticated concentrators are available to concentrate the solar energy 50 to 100 times, but they are best suited for commercial customers. However, such concentrators are too expensive and hence, are not recommended for domestic customers. A two-axis tracking solar array with 1 kW power rating and using compact fresnel lenses has concentrated solar energy 60 times. These lenses have optical efficiencies better than 85 percent and their grooves are turned inward to prevent dirt from accumulating on the array. The size and efficiency of the fresnel lens is dependent on the solar panel output power rating.

Concentrators are considered a "hot" price-reduction strategy by solar energy experts and are being developed in large quantities for commercial solar power applications.

Note solar concentrators can collect and focus more of the sun's energy falling on solar cells using compact parabolas, fresnel lenses, and optical lenses with other exotic configurations. Tests performed on these concentrators can boost the equivalent efficiency of standard silicon solar cells almost by 15 percent and of gallium arsenide solar cells by 20 percent. Low-cost concentrators are most effective in the regions where sunlight is most intense but least diffuse. Both tracking and nontracking show improvement of the solar cells. However, tracking concentrators exhibit higher solar performance over the nontracking ones, but are more expensive and complex.

Nontracking solar concentrators are relatively inexpensive, but they effectively multiply the surface area of a solar cell as much as 20 times or 20 suns. Nontracking concentrators generally use parabolic shaped elements to collect the extra photons and form them into an optical beam similar to a flashlight. An array of concentrators and equal number of solar cells are mounted on a lightweight rack. This rack tilts the array for best angle for a given site's latitude and the rack assembly can be manually adjusted according the seasons. The optical alignment involving sun, lens, and solar cells become critical beyond a concentration ratio of 20 suns or so and for a fixed-frame array due to the earth's rotation. Under these circumstances optimum tracking of sun is limited. Slightly higher concentrator ratios better than 20 are possible with a nontracking concentrator system comprising high-density strips of solar cells and parabolic reflectors, which can generate 100 peak watts.

Concentration ratios higher than 60 times are only possible with two-axis, single-section tracking concentrators using fresnel lenses, which can generate 1000 W of peak power [5]. A two-axis, three-sections tracking concentrator can generate a peak power of 300 W using a solar-powered microprocessor and stepping motors to follow the sun. This type of tracking concentrator can cost 15 cents per peak watt in volume quantities. Note the tracking speeds must be slow enough for the stepping motors to move the frame of the tracking concentrator compatible with sun motion, while the concentrator assembly is under microprocessor control. Note both the microprocessor and the stepping motors consume only a fraction of the solar power collected. Such microprocessor-controlled tracking concentrators have demonstrated concentration ratios better than 300 suns. However, such concentrator requires very accurate tracking and updating its position 30 times a minute [5].

1.6 Cost Estimates for Solar Modules, Panels, and Systems

It is of critical importance to understand the cost estimates and projections for solar modules and panels needed for construction of a solar power system. Cost estimates for the domestic solar installation will be different than that for a commercial office building or a discount store. It is important to mention that efficient design and low-cost installation of a solar energy system or a photovoltaic power source is essential before it can compete with the fossil-fuel power plants [4]. According to the solar scientists, if the sun provides the solar radiation 5 hours per day and operating over 365 days, the grid-connected solar energy will cost about 12 cents per unit or per kWh, which is slightly higher than 9 cents per unit usually charged by the utility companies involved in operating fossil fuel-based power plants. Utility companies' electricity rates are slightly lower for commercial customers compared to domestic customers. Solar module costs depend on the semiconductor materials, methods of connecting the solar cells in parallel and series configurations, encapsulation of finished cells, integration of cells into module, testing of each module prior to installation on the panel, module power output capability, and concentrator configuration for each module.

Solar cells are fabricated by the same process including the diffusion process that produces the transistors or integrated circuits. Cells are fabricated on round wafers, which are made by growing a cylindrical ingot of high-purity, single-crystal silicon. Wafers can be made in various diameters ranging typically from 3 to 5 in. An ohmic, current- collecting grid is deposited over the wafer face, which becomes one of the solar cell's output terminals and the other terminal is formed by the metal deposition on the backside of the cell. Output voltage polarity is determined by the direction on the p-n junction and the current flows in the forward direction.

These solar cells are arranged in parallel and series configurations using commercially available modules from various sources. Solar modules are available with various peak power ratings ranging from 50 W, 25 W, and 10 W. Such modules can be installed on solar panel arrays of appropriate dimensions. The solar arrays are protected by low loss and light weight transparent covers with minimum losses. These solar arrays are available with 1-, 10-, 25-, or 100-kW solar power capability.

Studies performed on various materials by the author reveal that the maximum theoretical conversion efficiency is 21 percent for silicon cells and 28 percent for gallium arsenide solar cells, excluding the losses due to dirt, metal-mask shadows, temperature, and I^2R term. In the case of silicon solar cell, unmatched spectral response reduces the available photons to 44 percent, the band-gap losses further reduce them to 29 percent, the fill factor reduces further to 27 percent, and finally the reflection effects and shadows bring down to a maximum theoretical efficiency of 21 percent for silicon solar cells, as illustrated in Figure 1.9. Solar cells made from thin films of silicon and gallium arsenide materials are getting more popular, because of higher conversion efficiencies and lower production costs. Solar energy experts predict worldwide market growth from such cells is expected to exceed $1.8 billion in 2008, $3.5 billion in 2010, $5.1 billion in 2012, $6.5 billion in 2014, and $8.1 billion in 2016.

Conversion efficiencies for commercially available solar arrays vary between 12 and 16 percent. Assuming the lowest array efficiency of 12 percent, the array converts the incident sunlight at the rate of 120 W/m² of the array area into usable electrical power. Assuming the array efficiency of 16 percent, the array converts the incident sunlight at the rate of 160 W/m² into usable electrical power. So the higher the solar array efficiency the higher will be the power conversion rate from solar to electrical energy. In other words, solar arrays with higher conversion rates will be found most cost effective for the consumers in converting the solar energy into electrical energy.

1.7 Solar Cell Performance Degradation and Failure Mechanisms in Solar Modules

Geographical locations and climatic conditions have a profound effect on the power output of the solar cells and modules. Peak power levels are generated only during the ideal conditions under the summer sun and clear skies. Furthermore, peak wattage only occurs at noon. During cloud cover, night darkness and the sun's acute angle of incidence in winter conditions, the average peak power cuts the annual amount of usable power down to 25 percent of the peak in the Phoenix area and down to 16 percent of the peak in the Boston area. Temperature takes a toll on the solar array performance. The solar array's design voltage of 15 V will drop to 13 V if the noontime temperature reaches 43°C. This drop of voltage amounts to a reduction of about 14 percent due only to temperature.

Properly interconnected and plotted solar cells can be expected to yield higher conversion efficiency. In fact, neither NASA nor JPL (the Jet Propulsion Laboratory) has been able to find measurable performance degradation or solar cell failures, even after several years of operation in space. However, back on earth, some reliability problems do occur, such as problems with encapsulating and potting materials in solar cell modules. The solar modules must be designed to be weatherproof, so that they can withstand day and night temperature cycling. The presence of moisture on the solar cell surface degrades the cell performance significantly due to losses by absorption and refraction. Moisture deposition on the cell surface also promotes the growth of organic substances, which effectively block the valuable sunlight. The ultraviolet rays of the sun darken the encapsulating material, leading to further reduction in the cell performance.

Sticky surfaces of modules collect dirt. To avoid this problem, module manufacturers must cover the module's face with clear plastic material or tempered glass to keep the module surface clean at all times. The module surface must be such that it should stay clean as much as possible under moderate rainy and snowy conditions. Cell interconnections within the solar module can become sources of failure mechanisms. To avoid this problem, reduction in interconnections should be made with minimum cost and complexity.

Solar arrays offer the solar energy while the sun shines, but not during the night. Therefore, proper power conditioning and storage facilities are required for any commercial solar-electric energy system just to maintain continuous operation during the night. Automobile-type storage batteries should be used as power buffers. In addition, owners of commercial solar power systems must keep a reasonable number of spare parts for the system they own such as solar cell modules, shunt regulators, blocking diodes to prevent shorting out the batteries at night, critical mounting hardware, and electrical cables to avoid emergency shut down of the solar generating system.

Several companies are involved in the design, development, and manufacturing of solar cells, solar modules, and solar panels. Anyone can contact these companies for specific information on solar energy related problems. Such companies include SolFocus, Sun Edison, Nano Solar, Sun Power, Sun Light, First Solar, Uni-Polar, Wurtz-Solar, and Mitsubishi. Relevant technical information on solar cells, modules, and panels is available on their websites to meet emergency shortage of critical components.

1.7.1 Solar Power Generation Cost Estimates

It is extremely difficult to provide exact cost estimates for photovoltaic power generation. The cost of solar power generation involves the cost of the solar module consisting of several solar cells, solar concentrator, inverter, panel design configuration, panel installation, and miscellaneous mechanical hardware. Recent published reports on solar power sources indicate that commercial use of PV power in the United States has increased from 100 to 500 kW in 1970 to 100 MW in 2005. The price for residential solar electricity was expected to go down from $90 per peak watt in 1970 to $15 per watt in 1980 to $1.5 per watt in 1990 to about 25 cents per watt in 2005. Energy experts believe that solar electricity cost will come down to 12 cents per watt by the year 2015. These cost estimates are contingent on high-volume deployment of low cost, high efficiency, silicon-thin-film solar cell and thin-film gallium arsenide solar cells with unique design architectures. Both the weight and cost reductions are possible using low cost silicon ribbons and sheets. Furthermore, more efficient solar cell manufacturing processes are needed to achieve these two objections. Higher solar cell efficiency can be expected from exotic materials such polycrystal silicon or nanotechnology-based materials with high absorption coefficients and V-groove multijunction (VGMJ) cell architectures. However, higher solar output power can be achieved through the use of optical concentrators, which can boost the power by a factor of 10 or 100 depending on the intensity of sunlight enhancement, but at the expense of higher costs.

1.7.2 Techniques for Optimization of PV Power Systems

High speed computers can play a key role in design configuration and optimization of PV power systems based on load factor, weather conditions, geometrical parameters, sun concentration factor, and availability of the solar cell materials with high light absorption capability. System configuration is known as "sizing," which normally includes a computer analysis of the installation site's latitude, longitude, mean temperature, and yearly insolation, which refers to the amount of the time the site is in the sun or "in-sol". The unit of measurement for insolation is equal to 1.62 mWh/cm². Armed with a description of the power load requirement and the circumstances surrounding the solar site, a computer can determine the most efficient solar generating system and associated components such as the number of solar modules and their series-parallel connections, the array's compass heading and the tilt angle, number of batteries required and their connections, and the projected system performance on a monthly basis. Computers can drive the procurement costs for flat-plate solar arrays as low as the process and materials will allow. However, the concentrators described below can accomplish optically what process and materials cannot do for large solar cell installation.

It is important to mention that a sunny bright summer day yields a peak energy distribution of about 1 kW/m² or 100 mW/cm² of the solar cell area. Just lying in the sunlight, a 2-in. diameter solar cell provides power of 0.25 W, 3-in. diameter provides power of 0.50 W, and 4-in. wafer diameter provides a power of 1 W. This indicates empirically that solar cell power output is proportional to the square of the cell diameter. Off course, these output power projections are without the use of optical concentrators. As stated earlier the use of optical concentrators or CPV technology will boost the output power by a factor 10 to 100 depending on the sun concentration factors. Deployment of optical concentrators will increase the installation cost slightly, but it will increase power output from the solar panel tremendously.

1.7.3 Techniques to Reduce Cell Cost and Improve Efficiency

Several countries have embraced various nonfossil energy sources with major emphasis on solar technology, because of greenhouse effects and the high cost of electrical energy generated using fossil fuels such as oil, gas, coal, and nuclear sources. Solar energy generation has been in high gear since 2000. Since 2005, the United States has produced more than 500 MW, Japan has produced more than 1550 MW and Spain has produced more than 65 MW of solar power. The U.S. solar output power represents less than 0.1 percent of the overall energy consumption [6]. Rapid acceptance by major countries is hindered because of higher costs and poor efficiencies of solar cells and modules. In brief, efficient photon harnessing, solar cell efficiency, and economical conversion into electrical energy remain technological challenges. According to solar energy experts, approximately 90 percent of the photovoltaic power generating systems now, and most likely into the foreseeable future, will continue to use silicon for the fabrication of solar cells. This material has many shortcomings. Several other technologies such as thin films, plastics, multifunction concentrators, and even nanotechnology-based quantum dots have their own strengths and weaknesses. Various issues related to cost and efficiency of solar cells and solar modules using organic and nanotechnology-based materials such as quantum dots, nanowires, nanocrystals, and carbon nanotubes (CNTs) are discussed later. The solar cell design concept using CNTs is illustrated in Figure 1.10.

1.7.3.1 Low Cost and Efficient Solar Cells

Traditional fossil-based electrical power generating sources are less expensive compared to photovoltaic power generating sources and thus relatively few countries have embraced solar power technology. However, high fuel prices, frequent political instability in several oil producing regions, and health concerns from carbon emissions have forced several countries to look for alternate, safe, reliable, and environmentally friendly energy sources. Solar technology offers such an alternate energy source, and this slowly maturing technology is poised to blossom in coming years into a significant PV electrical power source.

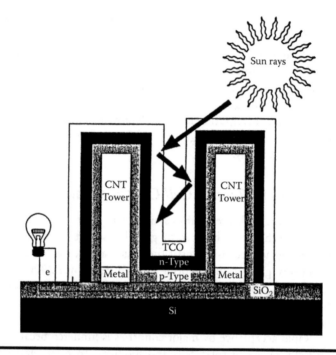

Figure 1.10 Photovoltaic cell design configuration using an array of carbon nanotubes capable of absorbing incoming photons from sunlight. (Data from Reference 6.)

It is important to point out that to achieve maximum efficiency the solar cell must be able to harvest the sun's energy through photosynthesis by squeezing solar energy out of every photon. Transient absorption spectroscopy can be used to measure the extremely fast protein movement in the reaction center and then match this movement to how the electrons are transferred in the center of the cell. Proteins within the reaction center can guide the electrons correctly in the semiconductor medium. Based on this photosynthesis theory, more efficient design of solar cells can be achieved to boost the performance of organic PV cells by incorporating polymeric solvents, which allow the cells to perform under changing sun conditions. Based on these facts, scientists are undertaking aggressive research and development activities on thin-film organic solar cells.

Most PV technologies operate under the same general principle. For example, each solar cell consists of two semiconductor materials (p-type and n-type), creating a positive and negative layer. When a photon ray from the sun strikes the cell with appropriate energy, electrons are knocked free and their movement from one layer to the other generates electricity. Existing material technologies exploit this phenomenon with various degrees of success or efficiency.

The performance of silicon PV cells has been improving at a faster rate. Ten years ago, its best efficiency was 16 percent, which is now close to 21 percent.

This means that roughly one-fifth of the sunlight hitting the PV cell is converted into electricity. PV cells using thin-film gallium arsenide have demonstrated maximum theoretical efficiencies close to 28 percent and PV cells using thin films of gallium arsenide with multijunction configurations have demonstrated efficiencies better than 30 percent. The latest research indicates that an approach involving multijunction concentrators has demonstrated conversion efficiency better than 40 percent [6], but it is too expensive for widespread use. PV cells using appropriate nanotechnology materials promise theoretical efficiencies exceeding 45 percent, but the implementation of nanotechnology requires significant research, development, and testing to demonstrate cost effectiveness. Since the cost of solar cells remains the major constraint, acceptance of solar power installations in the United States is too slow as demonstrated by the data shown in Figure 1.6.

Based on the 2005 data, electricity charges using PV energy sources varied from 23 to 25 cents per kilowatt hour or per unit for residential customers. For commercial customers, the unit cost was between 16 and 22 cents. According to the Solar America Initiative Program, which was created in 2006 by the U.S. Department of Energy to promote the use of solar energy, the unit prices will be cut by more than 50 percent. But the charges per unit for both customers are still too high. If the PV power suppliers can offer electricity for the residential customers between 8 and 10 cents per kilowatt hour and for commercial customers between 6 and 8 cents per kilowatt hour, then there will be widespread acceptance of solar energy technology. If this occurs, then one can expect significant reduction in greenhouse effects and dependency on fossil fuels. Electricity currently costs from 6 to 17 cents per kilowatt hour for residential customers and about 5 to 15 cents per kilowatt hour for commercial customers [6]. Reduction in the cost of PV systems is only possible through improvement in the solar cell efficiency and deployment of materials capable of providing higher absorption capability, improved reliability, and maximum output power levels with minimum cost and complexity.

1.7.3.2 Identification of Low Cost PV Cell Materials

Preliminary studies performed by the author on solar cell materials indicate that cells using thin films of plastic offer the lowest cost (a few pennies per kilowatt hour), which will be most appealing to more customers who are cost sensitive. In addition, plastic cells are much lighter, highly flexible, and compatible with esthetic integration into homes and even automobiles. Despite all these benefits of plastic solar cells, these cells are not cost effective because of ultra-low conversion efficiency close to 6 percent, compared to 20 percent for thin-film silicon cells, 28 percent for thin-film gallium arsenide cells, and close to 40 percent for multijunction concentrators. The ultra-low efficiency of plastic cells is due to the fact that the band of absorption in plastic medium is very narrow, allowing only a small portion of the incoming light on the cell surface. If a frequency conversion scheme can be integrated into their photovoltaic system, most of the incoming sunlight can be absorbed, leading to a significant improvement in cell efficiency. In other words, solar scientists think that a thermal annealing technique can be used to create a delicate network within a light-absorbing plastic medium. If the efficiency for the plastic solar cells can be improved better than 16 percent, then the plastic solar cells have great potential to be commercially viable for mass scale photovoltaic installation systems. Some research scientists think that certain organic polymers with small band gaps can absorb more photons, leading to higher conversion efficiency. Material scientists feel that currently no such material exists that can simultaneously satisfy three critical parameters, namely, band gap, high absorption capability, and ultra-low cost. Stanford University research scientists are currently working on the design and development of ultra-low cost organic material-based solar cells.

Scientists have been working for the last five years on solar cells using ternary compound semiconductor materials such as copper-indium-selenide (CIS), copper-gallium-selenide (CGS), and their multinary alloy copper-indium-gallium-selenide (CIGS). They found that solar cells made from a single crystal of CIS were very promising, but the complexity of the material appeared problematic as far as making thin films are concerned. The best laboratory efficiency for CIS cells are close to 19 percent. Solar cells fabricated using thin films of cadmium telluride (CdTe) did not exhibit conversion efficiencies better than 16 percent. Nanotechnology scientists reveal that solar cells made from nanotechnology-based materials such as carbon nanotubes (CNTs), quantum dots, and nanoparticles can provide theoretical conversion efficiencies better than 45 percent, but they suffer high fabrication costs and complex processes. Some of the processing techniques in designing solar cells using the ternary compound semiconductor materials, polymer organic materials, and nanotechnology-based materials are relatively very costly, but no meaningful data on their reliability and lifetime are available. It is important to mention that none of the above solar cells, except the silicon-based cells, can meet the minimum lifetime of 20 to 25 years. Potential materials for solar cell applications and their critical properties will be summarized later on in a separate chapter.

1.8 Summary

The history of the development of silicon photovoltaic cells is summarized, highlighting the latest research and development activities revealing performance data on various solar cells. Critical elements of a silicon solar cell including the p- and n-layer materials are identified. Reasons to justify the acceptance of solar energy for domestic and commercial customers are summarized. Critical design aspects and parameters for silicon solar cells are revealed. Potential applications of photovoltaic (PV) technology such as navigation aid sensors, radio relay stations, railroad communications networks, solar energy sources for domestic use, educational TV programs, and space vehicles are briefly discussed. Worldwide PV installation capacities are identified. Solar cell, solar modules, and solar array requirements for commercial and corporate buildings are summarized, with emphasis on cost, performance, and reliability. Impact of state and federal rebates and incentives on solar energy systems is discussed in great detail. Material requirements and their critical properties for silicon, gallium arsenide, ternary compound semiconductors, and nanotechnology-based solar cells are briefly discussed with emphasis on conversion efficiency and reliability. Scientists predict that solar cells using thin films with high absorption capability offer the highest conversion efficiencies. Material scientists predict that conversion efficiency can be expected as high as 45 percent for multijunction, three-dimensional nanotechnology-based solar cells using a specific material. Note nanotechnology-based materials include nanowires, nanodots, cabon nanotubes (CNTs), and quantum dots or nanocrystals. Plastic and polymer organic materials for solar cells are discussed briefly with emphasis on conversion efficiency. The production costs for plastic- and polymer organic-based solar cells are lowest, but their conversion efficiencies vary between 4 and 6 percent. Material scientists are working very hard to improve their conversion efficiencies. Fabrication requirements for solar modules and solar arrays are summarized. Concentrator technology for solar cells is discussed in great detail, because it offers higher conversion efficiencies and electrical power output levels for a given solar array size. Low-cost solar concentrator devices are described with emphasis on concentration factors and conversion efficiencies as a function of reflector size. Rough cost estimates for solar cells, solar panels, and solar power generating systems for domestic and commercial customers are provided. Failure mechanisms for solar cells and solar arrays are identified. Sources for solar cell performance degradation are discussed. Techniques to reduce solar cell cost and to enhance conversion efficiency are briefly discussed.

References

1. Dave Clark, R. Patel, et al. "Laser may bring the solar cell market," *Photonic Spectra*, June 2007, 54–58.
2. Sandra Upson. "Google corporate rooftops are the latest frontiers in solar cell energy generation," *IEEE Spectrum*, October 2007, 25–28.
3. Gary Conley. "Concentrated efforts," *SPIE Professional*, October 2007, 19–20.
4. Dick Hackmeister, Western Editor. "Solar cell technology," *Electronic Design* 26 (1977, December 20): 24–28.
5. Editor. "Photovoltaic cell advances—but slowly," *Electronic Design* 26 (1977, December 20): 24–28.
6. Michael A. Greenwood, News Editor. Solar technology: Seeking its day in the sun," *Photonic Spectra*, July 2007, 42–50.
7. W.L. Berks and Werner Luft. "Photovoltaic solar arrays for communication satellites," *Proceedings of IEEE* 59, No. 2, 263–270.

Chapter 2

Design Expressions and Critical Performance Parameters for Solar Cells

2.1 Introduction

This chapter will derive the design equations and identify critical performance parameters for solar cells using silicon and gallium arsenide materials. Silicon material is selected for fabrication of solar cells because of maturity of silicon and lower fabrication costs. Silicon comes in various forms, such as poly-silicon, single-crystal silicon, amorphous silicon, crystalline silicon, super-crystalline silicon, and ribbon-silicon. Crystalline silicon can be made either from monocrystalline silicon or multicrystalline silicon wafers. It is important to mention that more than 94 percent of solar cells are made from crystalline silicon semiconductor. As mentioned in the first chapter, the maximum theoretical conversion efficiency is 16 percent for silicon-based solar cells and 28 percent for gallium arsenide-based solar cells. Currently these two materials dominate the photovoltaic cell market, for the reasons cited above. Regardless of the materials used, the spectral response from these solar cells depends on the depth of the p-n junction, the absorption coefficient of the fabrication materials, cell junction area exposed to sunlight, and wavelength

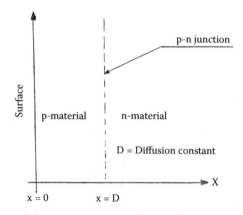

Figure 2.1 One-dimensional geometry for the theoretical model of a PV cell.

of the solar incident light. A solar cell is a semiconductor device that uses a p-n junction to convert solar energy directly into electricity. The magnitudes of the open-circuit voltage and short-circuit current are strictly dependent on the absorption capability of the materials used in the fabrication of the solar cells. The active surface of the cells currently in use consists of a thin layer of p-type silicon on top of a body of n-type material.

Solar cells are made with junction depths varying from 0.6 to 5.0 μm and can have smooth or rough surfaces. Existing response curves [1] indicate that in order to increase the short wavelength response at a wavelength less than 0.75 μm the junction should be made closer to the surface, while in order to increase the long wavelength response at wavelength greater than 0.75 μm should be made far below the surface. The effect of rough surface is to reduce the life time near the surface, thereby reducing the response to short wavelengths of the sunlight. A theoretical model of the cell shown in Figure 2.1 must be used to describe the mechanism involved in determining the shape of the response curves.

2.2 Spectral Response of Solar Cell Structure

The spectral response [1] of a solar cell is defined as the short-circuit current as a function of the wavelength of the incident light. This response is strictly dependent on the depth of the p-n junction and surface conditions. The silicon solar cell for laboratory investigations consists of a silicon slice with thickness ranging from 15 to 25 mil (1 mil = 1/1000 inch) and with resistivity of 0.5 ohm-cm. The surface irregularities must be removed for acceptable device performance. The fabrication of the device can be completed by depositing p- and n-type materials on the silicon slice. The contact to the n-material can be made by evaporating gold into the slice and the contact to the p-type material can be made by evaporating an aluminum

Table 2.1 Typical Electrical Properties of a Silicon Solar Cell as a Function of Diffusion Time and Junction Depth

Surface Conditions	Diffusion Time (min)	Junction Depth (μm)	Open-Circuit Voltage (V)	Short-Circuit Current (mA/cm²)
	6	0.6	0.52	15.5
Smooth	12	1.0	0.52	13.9
	18	1.35	0.51	11.3
	24	1.72	0.48	12.9
	6	0.60	0.32	10.5
	12	1.00	0.32	8.6
Rough	18	1.35	0.30	10.0
	24	1.71	0.40	9.4
Typical Cell	18	—	0.55	28.2

Note: The above parameters are obtained in brief sunlight conditions and are normalized to a light intensity of 1 kW/m².

dot into the slice. Since aluminum is itself an acceptor-type impurity in the silicon, the p-n junction is not destroyed but rather is simply moved deeper into the silicon slice in the vicinity of the dot.

The fabrication procedure is quite different for making commercial solar cells with typical junction depth of 2 μm to meet cost and yield specifications. Surface damage must be kept to a minimum during surface finishing and the surface must not exhibit reflections. Note the surfaces must be as clean as possible. The electrical characteristics of the cells will determine the spectral responses when exposed to sunlight with different wavelengths. Typical electrical characteristics of solar cells as a function of diffusion time, junction depth, open-circuit voltage, and short-circuit current are summarized in Table 2.1.

2.2.1 Impact of Spectral Response Parameters on Cell Performance

Critical spectral response [1] parameters, including the diffusion constant or coefficient, wavelength of the incident light, and the diffusion lengths in the p and n regions have significant effects on the reliability and performance of the solar cell.

A 10-μm diffusion length in the p region offers a lifetime of about 0.1 μsec in the n-material, which is considered a reasonable estimate. But for all junction depths, the values of L_n must be made larger for deeper junctions and smaller for shallower junctions, while the value of parameter L_p remains the same. This leads to the conclusion that a considerable reduction in lifetime as a result of contamination exists near the cell surface. However, as the junction progresses deeper into the material, the effect from surface contamination becomes less noticeable. This means that the average lifetime in the p-material must be longer for a deeper junction. In addition, the effect of surface overlapping will become less important for deeper junctions. Preliminary response calculations indicate that the relative response falls off faster for incident wavelengths less than 0.5 μm. This is due to the fact that as the wavelength approaches 0 and the absorption constant approaches infinity, the hole-electron pairs are generated near the surface in low-lifetime materials. Note deviation from the optimum wavelength values in the equal-energy spectrum will introduce a loss mechanism, leading to a reduction in the solar cell efficiency.

In addition, the effect of a reduction in lifetime in the vicinity of the surface due to surface damage or contamination is to reduce the short-wavelength response, with no effect on the incoming energy response at longer wavelengths. This means that it is important to characterize the principal mechanism in determining the shape of the response curves as a function of incident light wavelength.

2.3 Theoretical Model of the Silicon Solar Cell

A simple theoretical model shown in Figure 2.1 can predict the spectral response. The p-layer can be represented by a single lifeline. The fraction of the incident photons reflected from the surface is independent of the wavelength of the incident sunlight. Figure 2.1 indicates the geometry of the model and the coordinates assumed. For incident photons with energies exceeding the forbidden energy gap, the density of photons [1] in the silicon can be expressed as

$$N = [N_0 \exp{(\alpha x)}] \tag{2.1}$$

where N_0 is the surface density per in.^{2}, N is the photon density per in.^{2}, and α is the absorption coefficient. It is critical to mention that the density of the photons is proportional to the wavelength of the incident light on the cell surface, because the energy of an individual photon is proportional to the optical frequency.

Note that hole-electron pairs are created according to the negative of the rate change of the photon density and thus, the volume density of the hole-electron pairs created can be written as

$$N\,dx = [N_0 \times \exp{(-\alpha x)}\,dx] \tag{2.2}$$

where the parameter x varies from 0 to D and D is the diffusion constant.

The total number of carriers (Nc) crossing the p-n junction [1] can be given as

$$Nc = (N_0\, \alpha/\alpha - 1/L_n)\, [\exp(-D/L_n) - \exp(-D\,\alpha)] + [(N_0\, \alpha/\alpha + 1/L_p)\, \exp(D\,\alpha)] \quad (2.3)$$

where L_n is the diffusion length in the n region, L_p is the diffusion length in the p region, and α is the absorption coefficient, which is a function of wavelength and varies from junction material to material.

It is important to mention that the number of minority carriers (electrons) reaching the device junction is due to creation of hole-electron pairs in the p region, where the parameter x varies from 0 to D and $L = L_n$ as illustrated in Figure 2.1. Similarly, the number of minority carriers (holes) reaching the junction is due to creation of hole-electron pairs in the n region, where the parameter x varies from 0 to infinity and the diffusion length L is equal to L_p.

The intrinsic carrier concentration (Ni) in the p-n junction is dependent on the band gap energy (E_g) of the material and the product (kT/q) of the Boltzmann constant and temperature (T). Using the room temperature kT/q product of 0.0258, the intrinsic carrier concentration comes to 6.5×10^{-10} for a silicon cell and 1.38×10^{-12} for a GaAs cell.

2.3.1 Short-Circuit Current

The solar cell performance is strictly dependent on the short-circuit current (unit charge) in the device junction and is a function of wavelength (λ), absorption coefficient (α), diffusion constant (D) and diffusion length (L) in the p and n regions. The expression for the short-circuit current [1] can be written as

$$J_{sc} = (\lambda/1 - 1/\alpha\, L_n)\, [\exp(-D/L_n) - \exp(-D\alpha) + (\lambda/1 + 1/\alpha\, L_p)\exp(-D\alpha)] \quad (2.4)$$

Assuming a value of 0.1 for the absorption coefficient, 2 µm for parameter D, 0.5 µm for parameter L_n, and 10 µm for parameter L_p, and inserting these parameters into Equation (2.4), computed values of short-circuit current as a function of wavelength (α) have been obtained, which are summarized in Table 2.2.

Table 2.2 Computed Values of Short-Circuit Currents as a Function of Wavelength

Wavelength (λ), µm	Short-Circuit Current (mA)
0.5	10.4
0.6	12.3
0.7	14.4
0.8	16.5
0.9	18.5

It is important to point out that higher short-circuit current requires the maximum number of carriers reaching the device junction. In addition, it is necessary to maintain long diffusion lengths in order to keep the short-circuit current large.

2.4 Parametric Requirements for Optimum Performance of Solar Cell Devices

2.4.1 Introduction

This section of the chapter will define the parametric requirements capable of providing optimum performance of solar cells fabricated with silicon (Si) and gallium arsenide (GaAs) semiconductor materials. These two materials were selected because most of the current solar cells are fabricated using such materials. Furthermore, reliable spectral response data on p-n junction solar cells fabricated with silicon (Si) and GaAs materials are available as a function of relevant parameters involved. These data indicate that absorption behavior of GaAs solar cells is superior over the silicon-based cells [2] because of steep absorption edge, as illustrated in Figure 2.2. In addition, low effective masses and low densities of states in GaAs material yield higher collection efficiencies. Note the theoretical efficiency of a silicon solar cell is close to 16 percent compared to 28 percent for the GaAs devices, which may approach 32 percent if the carrier lifetimes stay well below those for Si devices. The latest research and development studies indicate that the efficiency of silicon devices is approaching a limit. These stated efficiencies will experience a reduction of 2 to 4 percent due to surface reflections and external contamination factors. Now attempts will be made to investigate the theoretical performance behavior of

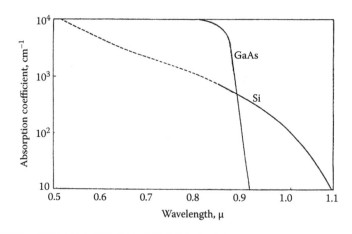

Figure 2.2 Absorption coefficient values for gallium arsenide (GaAs) and silicon materials as a function of solar spectral wavelength (μm).

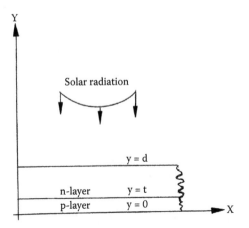

Figure 2.3   Diagram of a p-n junction solar cell.

these solar cells in order to find out the fundamental issues affecting the electrical performance levels of these devices.

2.4.2 Theory of Spectral Response of p-n Junction Devices [1]

The diagrammatic representation of a p-n junction solar cell is shown in Figure 2.3. The device junction consists of a thin p-layer from $y = 0$ to the junction at $y = t$, irradiated in the y direction at $y = 0$ so that I photons/sec are absorbed per unit area of the surface exposed to solar radiation. The n-layer extends from $y = t$ to $y = d$. The behavior of the p and n regions will be investigated ignoring the end effects, which means the problem will be treated as one dimensional.

2.4.2.1 Efficiency in the p Region for the Electrons

The continuity equation for the p region consisting of electrons [2] can be written as

$$dJ_e/dy = [(\Delta n/\tau_e) - KI \exp(-Ky)] \tag{2.5}$$

where τ_e (sec) is the lifetime of electrons in the p region, Δn is the concentrations of the photon carriers (cm^{-3}), K is the absorption coefficient (cm^{-1}), I is the photon flux (per cm^2 sec), y (cm) is the parameter indicating the depth or penetration in the surface, and J_e is the electron current (unit charge flow).

But the electron current in the absence of field component of the current compared with diffusion current is defined as

$$J_e = [D_e \, d \, \Delta n/dy] \tag{2.6}$$

where D_e is the diffusion constant for electrons, Δn is the concentrations of photon carriers (cm^{-3}), and d is the material layer thickness (μm). For short wavelengths and typical conditions, where parameter K is large, KL is much greater than one, and $\alpha < t/L < 1$, the expression for the short-circuit current (J_{sc}) can be written as

$$[J_{sc}/I]_e = [\cosh{(t/L)} + \alpha \sinh{(t/L)}]^{-1} - [\exp(-Kt)] \tag{2.7}$$

where α is the surface recombination parameter, t is the surface thickness, and L is the diffusion length. A typical value for t/L is 0.3, for α 0.1, t is between 1 and 3 μm, L is about 8 μm, K is between 0.3 and 1.0 μm, and Kl is greater than 2.

2.4.2.2 Sample Calculation for p-Region Efficiency

Since Equation (2.7) defines the ratio of saturation current to photon flux, it can be used to define the efficiency in the p region. Inserting the above values in the equation, one gets the efficiency in the p region as

$$[J_{sc}/I]_e = [\cosh{(0.3)} + 0.1 \sinh{(0.3)}]^{-1} + [\exp(-0.3 \times 3)]$$
$$= [1.045 + 0.1 \times 0.3045]^{-1} - [0.4066]$$
$$= [0.9298 - 0.4066] = [0.5232]$$
$$= [52.32] \text{ percent for } K = 0.3 \ \mu\text{m}$$
$$= [88] \text{ percent for } K = 1 \ \mu\text{m}$$

Plots for the p-region efficiency as a function of absorption coefficient (K) and normalized surface layer thickness (t/L) are shown in Figure 2.4. These plots indicate that the higher the absorption coefficient the higher the efficiency of the solar cell in the p region will be.

2.4.2.3 Efficiency in the n Region for the Holes

The continuity equation in the n region can written as

$$[dJ/dy]_h = [KI \exp(-Ky) - (\Delta p/t_h]] \tag{2.8}$$

$$[J_h] = [-D_h (\Delta p/dy)] \tag{2.9}$$

where D_h is the diffusion constant for the hole minority carriers, t_h is the lifetime for the holes in the n region, and K is the absorption coefficient for the holes. From these two equations, one gets the concentration of the photon-carriers as

$$[\Delta p] = [R \cosh(y/L_h) + S \sinh{(y/L_h)} - (G L_h/D_h) \exp(-Ky)] \tag{2.10}$$

where L_h is the hole diffusion length and is given as

$$L_h = [D_h \, \tau_h]^{0.5} \tag{2.11}$$

$$G = [KIL_h/\{KL_h)^2 - 1\}] \tag{2.12}$$

Using the boundary conditions, where $y = t$, $y = t + d$, $\Delta p = 0$ and a thick layer that offers exp $(-Kd) = 0$, coth $(d/L_h) = 1$, one gets the current at the junction at due to holes as

$$[J^h]_{Lh} = [P - G(KL_h - 1) \, \exp(-Kt)] \tag{2.13}$$

where P is a constant. When t is equal to zero, the p-layer is so thin that there can be no absorption in this layer. Under these assumptions, the expression for the short-circuit current (when $P = 0$) can be written as

$$J^h{}_{sc} = [K I L_h \, (1 + K L_h)] \tag{2.14}$$

Note the full current due to electrons can be obtained, if the absorption takes place well within the diffusion length in the n region. Furthermore, photo-holes produced in the n region will make an important contribution to the total current at longer wavelengths where the absorption is such that the p-layer is relatively transparent ($KL_h < 1$) but KL_h must be greater than unity to achieve photon-based efficiency. Thus, the total short-circuit current at the junction is the sum of currents in the n and p regions. However, it may be advantageous to have a surface n-layer with more mobile majority carriers and lower transverse resistance needed for higher collection efficiency. Since the relative magnitudes of L and L_h have very little effect on the efficiency, it will be convenient to use equal values for these diffusion lengths. Under these conditions and assumptions, the collection efficiency [2], which is defined as the ratio of J_{sc}/I, can be written as

$$[\eta_{coll}] =[J^k{}_{sc}] = KL \, [\alpha + KL - (1+\alpha) \, \exp(t/L - Kt)]/[(KL)^2 - 1)\{\cosh(t/L) + \alpha\sinh(t/L)\}] \tag{2.15}$$

This expression offers finite value when KL is equal to unity as a function of dimensional parameters KL and t/L. Note for maximum solar energy absorption, the p-layer must effectively absorb the solar energy within the diffusion length. The collection efficiency will be optimum for any absorption level such that the product KL is greater than one when t is given by

$$[Kt \, (\exp Kt - 1)] = [KL - 1] \tag{2.16}$$

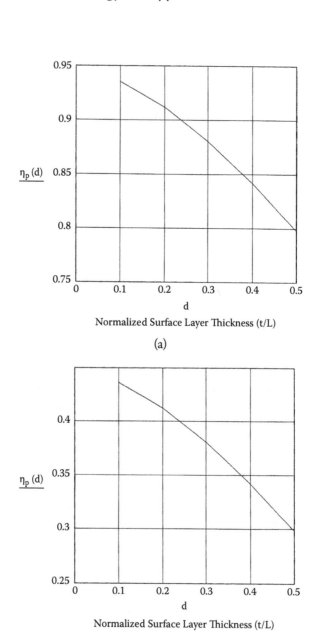

Figure 2.4 (a) p-region efficiency of a silicon solar cell as a function of the normalized surface thickness when the absorption constant $K = 1$ (a), 0.2 (b), 0.3 (c) and 0.5 (d). *Continued*

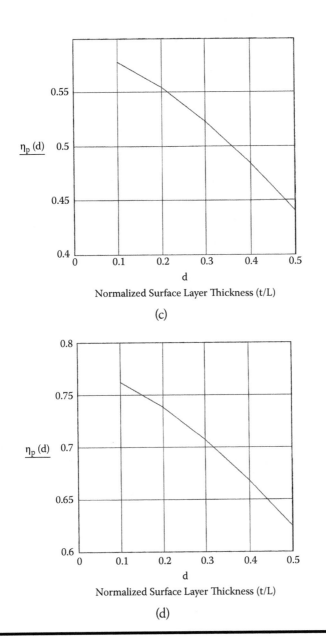

Normalized Surface Layer Thickness (t/L)

(c)

Normalized Surface Layer Thickness (t/L)

(d)

Figure 2.4 *Continued.*

Table 2.3 Computed Values of Parameter *Kt* as a Function of *KL*

KL	Kt
2	0.809
3	1.061
4	1.232
5	1.368
6	1.478
7	1.572
8	1.655

Inserting appropriate values of various parameters in Equation (2.8) through Equation (2.16), the collection efficiency can be computed as a function of dimensional parameters. The percentage collection efficiency plots as a function of various parameters involved are shown in Figure 2.3. Computations of collection efficiency require the values of *Kt*, which can be obtained using Equation (2.16). Computed values of *Kt* as a function of *KL* are summarized in Table 2.3. Note the parameter *L* stands for diffusion length, parameter *K* stands for the absorption coefficient, which varies from 0.3 to 1 μm, and the parameter *t* stands for the surface layer thickness, which varies from 1 to 3 μm.

Using the Mathcad 6.5 software program, collection efficiencies are calculated as a function of normalized surface layer thickness and *KL* and the efficiency curves are generated, which are illustrated in Figure 2.5. It is important to point out that the practical design of a solar cell requires surface layer thickness somewhat thicker than the optimum values in order to reduce the transverse resistance, which has an adverse impact on the collection efficiency of the cell.

2.4.3 Power Output of the Cell

The power output of a cell is the product of the open-circuit voltage (V_{oc}) and the junction current (J). The open-circuit voltage is written as

$$V_{oc} = [(kT/e) \log (1 + J_o/j_o)]$$
$$= [(kT/e) \log (J_o/j_o)], \text{ because the current ratio is much larger than 1} \quad (2.17)$$

where J_o is the total short-circuit current, j_o is the junction parameter, k is the Boltzmann constant (8.61×10^{-5}, eV/K), T is the absolute temperature (K) of the junction, and the quantity (kT/e) is equal to 0.258 or close to 0.26.

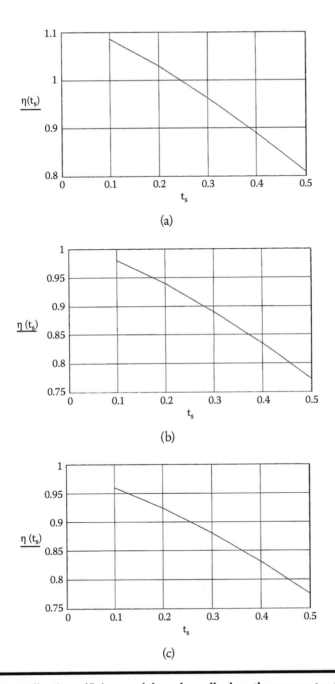

Figure 2.5 Collection efficiency of the solar cell when the parameter *KL* = 2 (a), 3 (b), 4 (c), 5 (d), 6 (e), 7 (f), and 8 (g). *Continued*

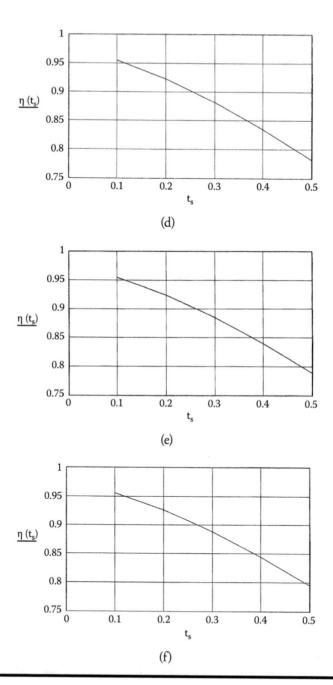

(d)

(e)

(f)

Figure 2.5 *Continued.*

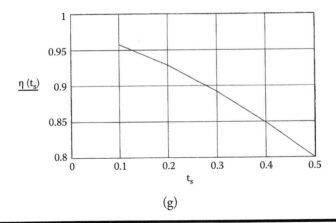

(g)

Figure 2.5 *Continued.*

The junction current J can be expressed as

$$J = [j_o \exp(eV/kT - 1)] - J_o \tag{2.18}$$

The optimum values of open-circuit voltage and junction current can be expressed as

$$V_{opt} = [r \, (kT/e)] \tag{2.19}$$

$$J_{opt} = [J_o \, r \, (r + 1)] \tag{2.20}$$

$$[(r + 1) \, e^r] = [J_o/j_o] \tag{2.21}$$

The parameter r can be computed by inserting values of J_o/j_o in Equation (2.21). This ratio has a value of 4.8×10^{12} for silicon solar cells, and 11×10^{17} and 4.7×10^{17} for normal GaAs solar cells and reversed GaAs solar cells, respectively. Inserting these values in Equation (2.21), one gets an r value of 26 for the Si devices, 38 for normal GaAs devices, and 37 for reversed GaAs devices. Once the vales of r are available, computed values for optimum power and conversion efficiency can be obtained.

The optimum power output of the cell, which is the product of optimum open-circuit voltage and optimum junction current, is defined by Equations (2.19) and (2.20), respectively, and is given as

$$P_{opt} = [(r - 1) \, (kT/e)J_o] \tag{2.22}$$

The conversion efficiency of the cell at optimum power output [2] can be given as

$$\eta = [(r - 1) \, (kT) \, (J_o/Q_o)]/E_g \tag{2.23}$$

Table 2.4 Collection Efficiencies (J_o/Q_o of Si and GaAs Solar Cells (Electrons per Solar Photon; Percent)

Dimensional Parameters (µm)	Silicon Cell	Gallium Arsenide Cell
$L = l = 8, t = 2$	37	—
$L = l = 8, t = 3$	38	—
$L = 10, l = 4, t = 2$	—	42
$L = 4, l = 10, t = 1.6$	—	39

where Q_o is the total rate of arrival of the solar photons in the semiconductor junction, which is also known as photon flux (cm^{-2} sec^{-1}), the product of kT is equal to 0.258 at room temperature, and E_g is the average energy of the photons, equal to 1.4 eV. Note for solar radiation outside the earth's atmosphere where the radiation intensity is 130 mW/cm^2, the photon flux (Q_o) is equal to 5.8×10^{17} photons per second per cm^2.

The ratio (J_o/Q_o) is also called the collection efficiency in electrons per solar photon and is dependent on diffusion length (L) for surface, diffusion length for bulk, and surface thickness (t). Under optimum matching conditions the solar cell is seen to behave as though each photon carrier reaches to the junction, thereby providing the incident solar energy to the external circuit. Typical values of collection efficiencies for silicon and GaAs devices as a function of various dimensional parameters in terms of electrons per solar photon are summarized in Table 2.4.

Collection efficiencies [4] for single-junction Si and GaAs solar cells as a function of wavelength of the incident solar energy are shown in Figure 2.3.

2.4.4 Theoretical Conversion Efficiencies of Single-Junction Si and GaAs Solar Cells

Maximum conversion efficiencies [2] for single-junction Si and GaAs solar cells can be computed by inserting the values of r parameters using Equation (2.21) and the collection efficiencies of the cells shown in Table 2.4 into Equation (2.23). Sample calculations for the Si and GaAs solar cells are provided for the convenience of readers.

$$[\eta]_{Si} = [(26 - 1) (0.0258) (37)/1.4]\%$$
$$= [17.05]\% \text{ for 37\% collection efficiency}$$
$$= [17.51]\% \text{ for 38\% collection efficiency}$$
$$[\eta]_{GaAs} = [(38 - 1) (0.0258) (42)/1.4]\% \text{ (normal case)}$$
$$= [28.64]\% \text{ for 42\% collection efficiency (normal case)}$$
$$= [(37 - 1) (0.0258) (39)/1.4]\% \text{ (reversed case)}$$
$$= [25.9]\% \text{ for 39\% collection efficiency}$$

Table 2.5 Theoretical Conversion Efficiencies for Single-Junction Si and GaAs Solar Cells as a Function of Collection Efficiency and Other Relevant Parameters

Solar Cell Type	r	Collection Efficiency (J_o/Q_o), Electrons/Photon	Conversion Efficiency (%)
Silicon cell	26	37	17.05
	27	38	17.5
GaAs cell (normal)	38	42	28.64
GaAs Cell (reversed)	37	39	25.90

These computations indicate that the higher the values of r and the collection efficiency the higher will be the theoretical conversion efficiencies for single-junction silicon and gallium arsenide solar cells. These are the maximum conversion efficiencies for single-junction solar cells, which will decrease by 3 to 5 percent due to various loss factors such as reflection at the front surface which may permit about 95 percent of the solar radiation to enter the semiconductor medium, imperfect electrical characteristics of the device, impurities in the semiconductor materials, and the contact efficiency. It is desirable to have a surface n-layer with more mobile carriers and lower transverse resistance. The conversion efficiency computations as a function of collection efficiency and the parameter r for both Si and GaAs devices are summarized in Table 2.5. Theoretical conversion efficiency and junction efficiency plots for silicon solar cells are shown in Figure 2.6 and Figure 2.7, respectively.

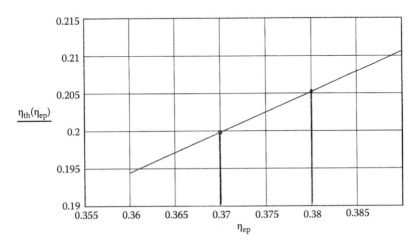

Figure 2.6 Theoretical conversion efficiency at optimum power output for a silicon solar cell.

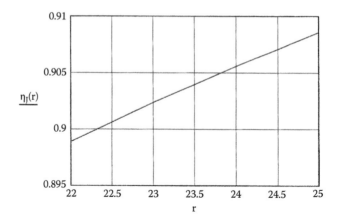

Figure 2.7   Theoretical junction efficiency at optimum power output for a silicon solar cell.

Theoretical conversion efficiencies of Si and GaAs improve with a rise in ambient temperatures. Estimated values of conversion efficiencies for Si and GaAs solar cells as a function of ambient temperature are summarized in Table 2.6.

Note these are maximum theoretical conversion efficiencies for single-junction solar cells, which will be reduced by 3 to 5 percent due to surface contamination, surface reflection, solar radiation aspect angle, and junction efficiency. Under minimum surface reflections, maximum photon energy enters the semiconductor junction with improved collection efficiency and higher conversion efficiency. Note surface doping has an impact on the collection efficiency. Heavier surface doping is required for the Si cells to obtain lower transverse resistance. The intrinsic carrier concentration also affects the conversion efficiency, because the carrier concentration varies exponentially with the energy gap (E_g) of the semiconductor material. Since the room temperature energy gap for GaAs is 1.42 eV compared to 1.1 eV for

Table 2.6   Impact of Ambient Temperature on Conversion Efficiency

		Efficiency (%) Si-Solar Cell			GaAs Solar Cell
Temperature (K)	kT/q	J_o/Q_o: 37	38	42	39
300	0.0258	16.23	18.26	28.86	26.09
325	0.0279	17.31	19.37	31.26	28.26
350	0.0301	18.94	21.31	33.67	30.44
400	0.0344	21.64	24.34	38.4	34.79

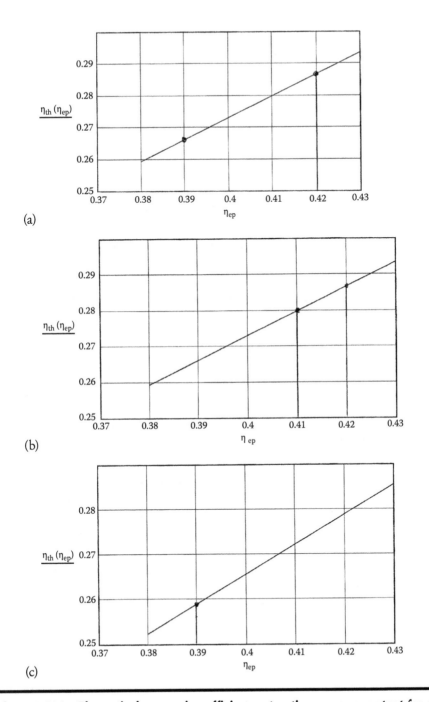

Figure 2.8(a) **Theoretical conversion efficiency at optimum power output for a normal GaAs solar cell (*r* = 38).**

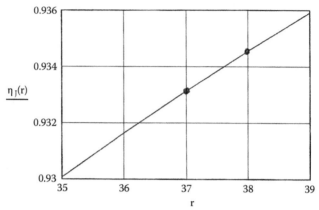

Note Typical Value of Parameter r is 38 for a Normal GaAs
Solar Cell and 37 for a Reversed GaAs Solar Cell

Figure 2.8(b) Junction efficiency for GaAs normal (r = 38) and GaAs reversed (r = 37) solar cells.

silicon, the intrinsic carrier concentration is roughly 10^5 times lower for the GaAs than Si material. That is why the conversion efficiency for a GaAs solar cell is higher than a Si solar cell.

2.4.4.1 Solar Module Power Conversion Efficiency as a Function of Open-Circuit Voltage, Short-Circuit Density, Sun Concentration Factor, and Form Factor (FF)

The power conversion efficiency of a solar module at a given temperature is dependent on the open-circuit voltage, short-circuit current and density, and form factor deployed by the solar module. Short-circuit current plots as a function of absorption coefficient are illustrated in Figure 2.9. Typical conversion efficiency of commercial modules is roughly 2 to 3 percent lower than the conversion efficiency of the solar cells. The solar module efficiency [3] as a function of short-circuit current, open-circuit voltage, fill factor, and input power from the solar spectrum can be computed using the following expression:

$$\eta_{mod} = [I_{sc}\, V_{oc}\, FF/P_{in}] \qquad (2.24)$$

where I_{sc} is the short-circuit current, V_{oc} the open-circuit voltage, P_{in} the input power from the solar spectrum, which is 0.135 W at AMO (air mass optimum) equal to one, and FF is the form factor used in the design of the solar module. Note the open-circuit voltage is the voltage at various sun concentration factors

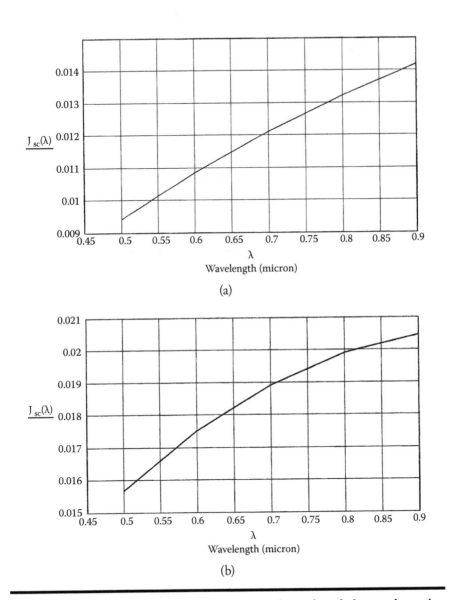

Figure 2.9   Short-circuit current as a function of wavelength for an absorption constant of 0.05 (a), 0.1 (b), and 0.2 (c). *Continued*

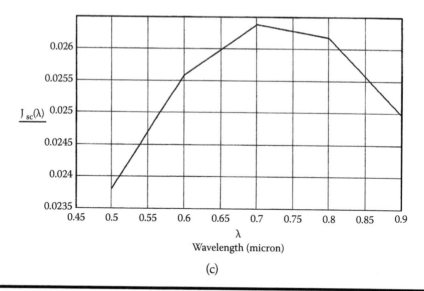

(c)

Figure 2.9   Continued.

as illustrated in Table 2.7. The power conversion efficiency data summarized in Table 2.7 are valid for silicon-based solar modules only. The power conversion efficiency figures will be 3 to 5 percent higher than those specified for silicon solar modules. Inserting values of various parameters in Equation (2.24), solar power conversion efficiencies are calculated, which are summarized in Table 2.7.

2.4.4.2 Maximum Output Power Density at 1 AMO and 300 K Temperature

The maximum output power density from a solar cell [2] is a function of short-circuit current density (J_{sc}), open-circuit voltage (V_{oc}) and form factor (FF) and can be given as

$$P_{max} = [J_{sc}\ V_{oc}\ FF]\ W/cm^2 \tag{2.25}$$

Computed values of maximum output power density as a function of various parameters are summarized in Table 2.8.

2.4.5 Optimum Open-Circuit Voltage for Single-Junction Solar Cells

Optimum open-circuit voltage for single-junction solar cells can be computed by inserting appropriate values of the parameter r shown in Table 2.5 into Equation (2.19).

Table 2.7 Solar Module Conversion Efficiency as a Function of Open-Circuit Voltage (V_{oc}), Short-Circuit Current (I_{sc}) and Form Factor (FF) (Percent)

Sun Factor F_{sun}	V_{oc} (V)	Conversion Efficiency at 300 K		
		$I_{sc} = 0.025$ A	$I_{sc} = 0.030$ A	$I_{sc} = 0.035$
		FF: 0.7/0.8	0.7/0.8	0.7/0.8
1	1.01	13.14/14.46	15.71/17.95	18.33/20.95
10	1.07	13.87/15.85	16.64/19.02	19.42/22.19
100	1.13	14.65/16.74	17.58/20.09	20.51/2.44

Table 2.8 Maximum Output Power Density as a Function of Various Parameters (W/cm²)

J_{sc} (W/cm²)	V_{oc} (V)	FF = 0.7	FF = 0.8
0.025	1.01	0.0177	0.0202
0.030	1.07	0.0229	0.0257
0.035	1.13	0.0277	0.0316

$$V_{oc} = [(26)\ (0.0258)]$$
$$= [0.671]\ \text{V for Si devices}$$
$$= [(38)\ (0.0258)]$$
$$= [0.980]\ \text{V for "normal" GaAs devices}$$
$$= [(37)\ (0.0258)]$$
$$= [0.956]\ \text{V for "reversed" GaAs devices}$$

Note these calculations do not take into account the effects of the sun concentration factor, which is provided by the solar concentrator [4], temperature, environments, diffusion limitations, and fill factor.

2.4.5.1 Open-Circuit Voltage for p-n Junction Devices in Diffusion Limited Cases

It is possible to achieve maximum open-circuit voltage with p-junction devices in diffusion limited cases [2] and the expression for the maximum value can be written as

$$V_{oc} = [n\ (kT/q)]\ [\log_e\ (J_{sc}/J_{sat} + 1)] \tag{2.26}$$

where J_{sc} is the short-circuit current density, J_{sat} is the saturation current density (which is sometimes referred to as J_o in the literature), n is a constant with a maximum value of 1, and the quantity (kT/q) is equal to 0.0258 for room temperature (300 K). Since the short-circuit current density to saturation current density ratio is much greater than one, the above equation is reduced to

$$V_{oc} = [(0.0258) \log_e (J_{sc}/J_{sat}] \qquad (2.27)$$

Assuming a short-circuit current density of 0.025 A/cm² and saturation current density of 1.3×10^{-25} A/cm², the open-circuit voltage under diffusion limited condition is equal to

$$V_{oc} = [(0.0258) \log_e (0.025/1.3 \times 10^{-25})] = [(0.0258)(53.611)$$
$$= [1.383] \text{ V at } 25 \text{ mA/cm}^2$$

Open-circuit voltage plots for Si and GaAs devices are shown in Figure 2.10 and Figure 2.11, respectively. Plots of open-circuit voltage and conversion efficiency as a function of sun concentration factor are displayed in Figure 2.12 and Figure 2.13, respectively.

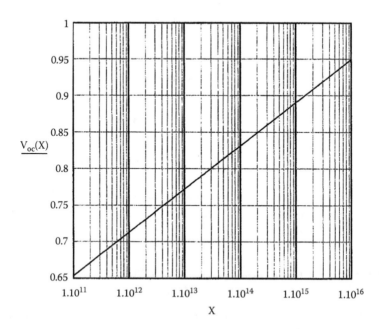

Figure 2.10 **Open-circuit voltage for a silicon solar cell as a function of photon current density.**

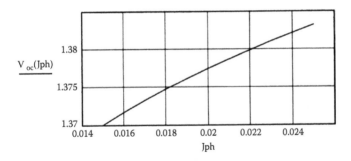

Figure 2.11 **Open-circuit voltage for GaAs solar cells as a function of photon current density.**

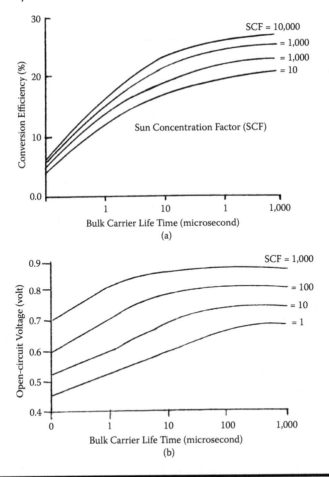

Figure 2.12 **Performance of a VGMJ silicon solar cell. (a) Conversion efficiency and (b) open-circuit voltage as a function of lifetime.**

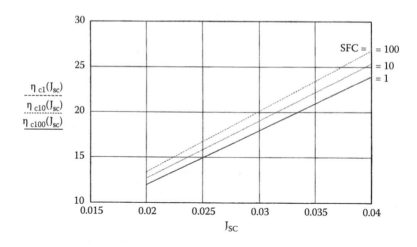

Figure 2.13 Conversion efficiency for a GaAs solar cell as a function of short-circuit current and solar concentration factor (SCF).

Table 2.9 Open-Circuit Voltage as a Function of Concentration Factor and Ambient Temperature (V)

Sun Concentration Factor	Temperature (K)		
	300	*400*	*500*
1	1.013	0.812	0.605
10	1.074	0.892	0.705
100	1.134	0.994	0.813

2.4.5.2 Open-Circuit Voltage as a Function of Sun Concentration Factor and Temperature

In actual practice, the open-circuit voltage is dependent on the sun concentration factor [4] and the ambient temperature of the surface. The published literature reveals that temperature has significant impact on the open-circuit voltage due to the sun concentration factor. Estimated values of voltage as a function of concentration factor and ambient temperature are summarized in Table 2.9.

2.5 Overall Conversion Efficiency of Solar Cells

The overall or the net conversion efficiency of a solar cell is dependent on several factors or issues, such as collection efficiency, junction efficiency, atmospheric

Table 2.10   Junction Efficiency Computations for Si and GaAs Solar Cells (Percent)

r	Junction Efficiency for Si Solar Cells	r	Junction Efficiency for GaAs Solar Cells
24	84.50	36	88.36
25	84.93	37	88.59 (reversed device)
26	85.33	38	88.82 (normal device)
27	85.72 (correct value)	39	89.09

absorption, reflection efficiency, contact losses, and solar aspect angle. The effects of collection efficiency, junction efficiency, and atmospheric absorption have already been discussed. The effects of reflection efficiency and solar aspect angle require detailed discussion. The impact of solar aspect angle on conversion efficiency will be discussed in a separate chapter.

2.5.1 Junction Efficiency

The junction efficiency of a solar cell is strictly dependent on the junction geometric and dimensional parameters and the junction materials. Note that the parameter r, which was previously discussed, indicates the junction material. Its value is 27 for silicon solar cells and 38 for GaAs (normal) solar cell and 37 for GaAs (reversed) solar cells. These values of r have been specified previously. The expression for the junction efficiency [2] can be written as

$$J_{jun} = [(r-1)/r + \log_e (r+1)] \tag{2.28}$$

Junction efficiency is computed for various values of r just to find out how further improvement can be achieved in the junction efficiency. Using Equation (2.28), computed values as a function of r for both silicon and gallium arsenide solar cells are obtained and are summarized in Table 2.10.

The comments in parentheses indicate the assigned value of parameter r. These computed values indicate that higher values of parameter r yield slightly higher junction efficiencies.

2.5.2 Contact Efficiency

Higher contact efficiency is essential to reduce the contact losses. Low resistance contacts can yield contact efficiencies in the range of 97 to 98 percent. Therefore, it is reasonable to assume a contact efficiency close to 98 percent for rough calculations.

Table 2.11 Net Conversion Efficiencies or the Overall Conversion Efficiencies for Si and GaAs Solar Cells

Product of Four Efficiencies	Silicon Cell (r = 27)	GaAs Cell	
		(r = 38)	(r =37)
0.75355/0.78557/0.78353	13.19	22.50	20.29
Cell type:	Normal	Normal	Reversed

Note: These overall conversion efficiency calculations use the combined efficiency product of 0.75355 when $r = 27$, 0.78557 when $r = 38$, and 0.78353 when $r = 37$. In the case of GaAs solar cell (normal) the net conversion efficiency is the highest among single-junction solar cells.

2.5.3 Absorption Efficiency

The ability of the cell surface to absorb the solar radiation energy plays a key role in determining the optimum conversion efficiency. In brief, the ability of the solar cell surface to absorb the solar energy over a wide spectral region is a key factor in determining the overall conversion efficiency. Its typical value of 95 percent is not unreasonable under optimum solar aspect angles.

2.5.4 Reflection Efficiency

Reflection losses from the solar module surface have been reported between 4 to 6 percent under the worst operating environments. Therefore, a reflection efficiency of 95 percent seems very reasonable, provided reflections from all sources are kept to a minimum.

2.5.5 Overall Theoretical or Net Conversion Efficiencies of Si and GaAs Solar Cells

It is important to mention that the estimation of overall theoretical or the net conversion efficiency must take into account the theoretical conversion efficiency, junction efficiency, contact efficiency, absorption efficiency and reflection efficiency. Assuming the various efficiencies specified above, the overall theoretical or the net conversion efficiencies for single-junction Si and GaAs are summarized in Table 2.11.

2.6 Critical Design and Performance Parameters for Silicon and Gallium Arsenide Solar Cells

Design and development engineers and scientists must have a clear understanding of the critical design and performance parameters besides the basic knowledge

Table 2.12   Critical Design and Performance Parameters for Single-Junction Silicon and Gallium Arsenide Solar Cells

Parameter	Silicon Cell	GaAs Cell (Normal)	GaAs Cell (Reversed)
Energy gap (eV)	1.1	1.42	1.42
Surface doping (p)	3×10^{18}	5×10^{17}	5×10^{17}
Bulk doping (n)	10^{18}	5×10^{17}	5×10^{17}
Surface thickness	2–3 μm	2 μm	1.6 μm
Diffusion constant (bulk) cm²/sec	7.5	7.5	60
Diffusion constant (surface) cm²/sec	25	60	7.5
Diffusion length, μm			
Surface	9	10	4
Bulk	9	4	10
n_i^2, cm^{-6}	4.1×10^{19}	0.4×10^{13}	0.4×10^{13}
j_o	4.5×10^5	0.22	0.48
j_o/n_i^2	1.1×10^{-14}	5.5×10^{-14}	12×10^{-14}

on solar spectrum and solar energy levels in different spectral regions and surface doping levels needed to reduce the transverse. Critical design and performance parameters for silicon and gallium arsenide solar cells include the energy gap in the material, theoretical efficiency limit, surface doping level, surface thickness, bulk doping level, diffusion constants (D) in surface and bulk, diffusion lengths (L) in surface and bulk, junction parameter (j_o), transverse resistance, saturation current, short-circuit current, open-circuit voltage, conversion efficiency, collection efficiency (electrons per solar photon), lifetime of the material, intrinsic carrier concentration (n_i^2), absorption efficiency, contact efficiency, junction efficiency, reflection efficiency, and (j_o/n_i^2). The most important design parameters are summarized in Table 2.12.

2.7 Solar Cell Design Guidelines and Optimum Performance Requirements

The most efficient design of solar cells requires preliminary review of potential design configurations and selection of the most promising cell design. Surface conditions and dimensional parameters must be compatible with the optimum performance requirements. Detailed analysis of the photoelectric processes in p-n junctions is necessary to meet specific conversion efficiency goals. It is important to mention that for any given absorption constant there is an optimum thickness of the surface layer. The optimum thickness is roughly the reciprocal of the absorption constant, averaged over the effective wideband solar spectrum. "Gridded" structures must be selected to reduce the transverse current, which adversely affects the conversion efficiency, regardless of the semiconductor material used in the fabrication of the devices. GaAs solar cells with the same dimensional parameters offer high conversion efficiencies over silicon devices. Improvement in conversion efficiency for silicon cells requires extensive research activities with respect to doping techniques, fabrication of sharp junctions, ultra-low resistance contacts, and reflectivity and surface recombination.

The fundamental performance advantages of the GaAs material over Si are:

- GaAs has a very sharp absorption edge over silicon, as illustrated in Figure 2.4, which allows direct transitions from valence to conduction band.
- Activation energy of GaAs material is significantly higher compared to that of Si.
- Lower effective masses in GaAs yield low-density states.
- Very high mobility in GaAs offers low transverse resistance in case of GaAs "reversed" solar cell configuration.
- The minority-carrier concentration for a given doping level is roughly 10^7 lower in GaAs material compared to that in silicon.
- In general, a well-designed GaAs solar cell offers slightly higher short-circuit current and about 50 percent higher voltage over the silicon cell.
- The intrinsic carrier-concentration of GaAs has a net advantage of $10^7{:}1$ over silicon.

The only fundamental disadvantage of GaAs is that it has a shorter lifetime than silicon.

2.8 Summary

The fundamental requirements for single-junction silicon and gallium arsenide solar cells are established in this chapter, with emphasis on high performance and minimum fabrication costs. Ternary compound semiconductor materials such as amorphous silicon (a-Si), cadmium telluride (CdTe), copper indium di-selenide (CIS), and copper-indium gallium di-selenide (CIGS) are being considered in the design and development of solar cells. Material research studies are being directed to develop organic materials for possible applications in solar cell fabrication. However, conversion efficiencies of organic solar cells at this time vary between 3 and 5 percent. Research efforts are underway at Stanford University, Palo Alto,

California to improve the conversion efficiency of organic solar cells. Organic solar cells will be discussed in a separate chapter. Computations and plots for the critical performance parameters, namely, the open-circuit voltage and short-circuit current as a function of surface conditions, photon current density, diffusion time, and junction depth are provided. A theoretical model for a silicon solar cell is described identifying the critical dimensional parameters. Fabrication processes and parameters for optimum performance are discussed. Parameters affecting efficiencies in the p region for electrons and n region for holes are identified. The power output equation from a solar cell is derived. Computed values and plots of collection efficiency (electrons per solar photon), conversion efficiency of the solar cell in the p region, junction efficiency and total conversion efficiency (percent) for both silicon and gallium arsenide solar cells as a function of dimensional parameters are provided for the benefit of solar cell designers. The impact of temperature, spectral absorption, sun concentration factor and fill factor on conversion efficiency is discussed. Techniques to enhance solar module conversion efficiency are identified. Magnitudes of open-circuit voltage for p-n junction devices in diffusion limiting cases are summarized. Overall or net conversion efficiency computations of single-junction solar cells, including the contributions from theoretical conversion efficiency, collection efficiency, junction efficiency, absorption efficiency, and reflection efficiency are provided. Critical design aspects, performance parameters, and major advantages of gallium arsenide single-junction solar cells are identified, with emphasis on conversion efficiency, reliability, and fabrication cost. Higher conversion efficiencies are possible using a solar concentrator with exotic geometrical configurations [4].

References

1. L.M. Terman. "Spectral response of solar cell structures," *Solid State Electronics* 2 (1961): 1–7.
2. T.S. Moss. "The potentials of silicon and gallium arsenide solar batteries," *Solid State Electronics* 2 (1961): 222–231.
3. A.R. Jha. "Solar heating and cooling of residential homes using solar cells." Synopsis. Jha Technical Consulting Services, Cerritos, CA.
4. A.R. Jha. "Technical report and cost estimate for a solar electric power system using solar cells." Jha Technical Consulting Services, Cerritos, CA.

Chapter 3

Classification of Solar Cells Based on Performance, Design Complexity, and Manufacturing Costs

3.1 Introduction

This chapter describes the classification of solar cells based on performance, design complexity, and manufacturing costs. For commercial applications, it is absolutely necessary to reduce the cost of high efficiency, which will meet the affordability criteria. However, high-efficiency solar cells based on cost-effectiveness require thinner semiconductor wafers (less than 200 nm) and replacement of current costly photolithography processes with low-cost, simple processing techniques. Currently high-efficiency silicon solar cells have been fabricated using a number of expensive photolithography processes involving thick silicon wafers. Studies performed by the author on low-cost solar technology indicate that passivated emitter and rear cell (PERC) technology and V-groove multijunction (VGMJ) solar cell technology offer design flexibility, high efficiency, and minimum cost of a solar energy system. The studies further indicate the PERC structure shows great potential for commercialization of high-efficiency silicon solar cells, based on thinner silicon wafers with a simplified process. Achieving cells exceeding 20 percent efficiency requires good passivation of both cell surfaces with high-quality dielectric layers such as SiO_2 and SiN_3. Various methods such as plasma etching, laser ablation, and mechanical abrasion have been investigated to pattern the rear contact on the dielectric layer instead of an expensive photolithographic process. A thermally grown silicon oxide layer can be used as the dielectric passivation layer in PERC solar cells, which offers lower spreading resistance and minimum contact resistance. High-efficiency solar cell design configurations require brief review of performance and fabrication costs of several solar devices, such as PERC solar cells, amorphous silicon (a-Si) cells, CIS solar cells, CIGS solar cells, CdTe solar cells, CdTe-CdS solar cells, nanocrystal (nc) dye-sensitized solar cells, CdHgTe solar cells, tendon-junction solar cells, VGMJ solar cells, silicon point-contact concentrator solar cells, Schottky-barrier solar cells (SBSCs), and multi-quantum-well (MQW) solar cells. It is important to mention that some cells will offer high conversion efficiencies. Some solar cells will be fabricated with minimum costs. Every effort will be made to identify solar cells capable of providing high efficiency, improved reliability in terms of operational hours, and minimum fabrication or manufacturing costs. In other words, solar cells capable of meeting the above three design criteria will be discussed in detail, while others will be discussed briefly.

3.2 Identification of Design Aspects and Critical Design Parameters for Low-Cost, High-Efficiency Solar Cells

Critical design aspects or fundamental design parameters for a low-cost, high-efficiency solar cell are identified below. Preliminary studies performed by the author on solar cell technology reveal that contact resistance plays a key role in the conversion efficiency of the solar cell, regardless of device type. Contact resistance losses occur at the interface between the solar cell and the metal contact. Contact resistance is the most critical performance parameter of the solar cell. Major performance degradation from the increased series resistance is due to the contact resistance that will ultimately degrade the conversion efficiency of the solar cell, regardless of the solar cell design.

For silicon-based cells, the fundamental recombination limit, the effects of diffusion gradients, resulting electric field loss, and contact insertion loss can degrade conversion efficiency of the cell. It is important to point out that the contact surface integrity is of critical importance, if longevity or ultra-long operational duration is the principal design requirement.

For point-contact solar cell devices, device geometrical parameters, surface recombination velocity, bulk lifetime, idealized thermodynamic limit, and emitter saturation currents play key roles, if high efficiency and ultra-high reliability are the fundamental design requirements.

For low-cost, high-efficiency PERC solar cells, fast processing time, mass production capability, low processing cost and low-cost, efficient patterning scriber technology are important. The photolithography process is the most expensive process and must be avoided, if minimum fabrication cost is the principal design requirement.

Fundamental design requirements for high-efficiency solar cells include good passivation of both cell surfaces with high-quality dielectric layers such as silicon oxide and silicon nitride, thin wafers, and mechanical abrasion technology to pattern the rear contact on the dielectric layers instead of an expensive photolithography process. The mechanical scriber process [1] shows great potential for low-cost commercial solar power module applications. Now some potential solar cells capable of providing both the low cost and high efficiency will be described in more detail.

3.3 Description of Potential Low-Cost, High-Efficiency Cells

Only solar cells capable of providing low manufacturing costs and high conversion efficiencies will be described in detail here, with particular emphasis on high reliability or long operating duration, which are the principal requirements of solar cells for implementation in commercial solar power systems.

3.3.1 Low-Cost, High-Efficiency Passivated Emitter and Rear Cell (PERC) Devices

Three distinct techniques are available for rear contact patterns on the silicon wafers used in the fabrication of PERC devices. The three rear pattern types include dot pattern using photolithography technique, dashed line pattern with mechanical scriber (MS), and line pattern involving MS. Solar cell performance parameters for three types of rear contact patterns on a 0.5-ohm-resistivity, p-type silicon wafer with cell area of 4 cm² are summarized in Table 3.1.

The data indicates that both high open-circuit voltage (V_{oc}) and high conversion efficiency require low-contact resistance and high fill factor (FF). The data further

Table 3.1 PERC Solar Cell Performance Parameters for Three Distinct Rear Pattern Types

Rear Pattern Type	Series Resistance (Ω)	V_{oc} (mV)	FF	Efficiency (Percent)
Dot pattern using photolithography	0.125	666	0.807	19.98
Dashed line pattern with mechanical scriber (MS)	0.213	646	0.795	19.42
Line pattern with MS	0.346	655	0.710	18.27

indicates that dot pattern with photolithography offers the highest conversion efficiency, but at higher fabrication cost. It is quite evident from this data that the dashed line pattern using MS technology offers fairly high efficiency, but at minimum cost. Reliability data on PERC devices are currently not available, because actual reliability data collection requires performance tests for several devices and over very long durations. However, reliability estimates can be achieved on these devices using theoretical reliability models.

There are two important design issues facing the PERC solar cell fabrication, namely the rear contact structure and the mechanical-scribing (MS) process. It is important to point out that the contact resistance is negligible in a PERC cell compared to the spreading resistance for optimizing the back contact spacing. Preliminary review of these three rear patterns indicates that the rear contact area fraction is 0.8 percent for a dot pattern using the photolithography compared to 0.9 percent for a dashed line pattern with mechanical scriber, which reveals a very small difference. However, the fabrication cost is the lowest in the latter case.

3.3.2 Mechanical Scribing Process for Fabrication of PERC Devices

The mechanical scribing process is depicted in Figure 3.1. This scribing process involves scribing damage, scribing damage etching, and removal of overhanging silicon oxide layer. The scribing tip of the mechanical scriber can be made from an artificial diamond to reduce the tooling cost. The minimum scribing width depends on the shape and geometrical configuration of the tip and typically varies between 5 and 10 μm. The maximum scribing velocity should be limited to 20 m/sec to achieve optimum performance. Uniform buffer spring force must be used on the wafer surface to scribe a uniform, very shallow depth over the entire silicon wafer surface with minimum surface defects. If a mechanical scriber has three scribing tips, it is possible to scribe three wafers simultaneously. The MS can be easily modified to attach more scribing tips for mass production of the devices, thereby allowing work on several devices simultaneously with minimum costs. Damage to the surface, if any, can be removed using KOH solution after mechanical scribing.

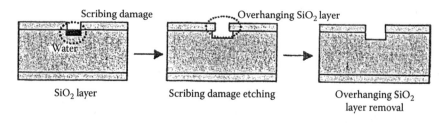

Figure 3.1 Mechanical scribing process used in the fabrication of PERC solar cells.

The overhanging silicon layers can be removed using HF solution. The overhanging layers prevent uniform deposition into the rear pattern when metals are evaporated. This rear processing scheme involves two etching steps.

3.3.3 Fabrication Steps

Specific details on the mechanical scribing process are illustrated in Figure 3.1. The inverted pyramid structures on the silicon wafers with front texturing are shown in Figure 3.2. A homogeneous n^+ emitter diffusion process is carried out at the front surface. Thermal oxidation is involved for metallization masking, passivation layers, and antireflection coatings. Rear contact is patterned using the MS process, while the front metal contact can be patterned with a photolithographic technique. Deposition of front metal contact and rear metal contact can be done with an electronic-beam evaporation technique. Silver plating of the contact layer should

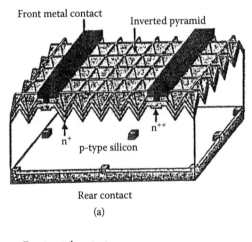

(a)

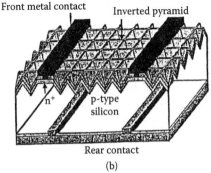

(b)

Figure 3.2 Structural details for (a) a conventional PERC solar cell and (b) a mechanical scriber-PERC (MS-PERC) solar cell.

be done to reduce contact loss. Structural details on conventional PERC solar cell and MS-PERC solar cell are shown on Figure 3.2. In a conventional PERC device, a selective emitter must be formed at the front side and the rear silicon oxide layers, and patterning can be done using photolithography. For an MS-PERC device, patterning must be accomplished using a mechanical scriber.

3.3.4 Performance Levels of PERC and MS-PERC Cells

Critical performance parameters of these solar cells are summarized in Table 3.1. The best efficiency for the conventional PERC solar cell with dot pattern is about 19.98 percent, while for the MS-PERC device with dashed line pattern is 19.42 percent. The conversion efficiency of the MS-PERC device is slightly lower than that of a conventional PERC cell due to lower open-circuit voltage. There are two possibilities for low voltage. The first one is poor rear contact of MS-PERC cells because of insufficient removal of mechanical damage or overhanging oxide layers. The second is due to a nonoptimized emitter diffusion process. If the diffusion process is optimized, the open-circuit voltage will increase higher than 670 mV and the conversion efficiency of MS-PERC devices will exceed 20 percent or higher. The mechanical scribing process is very effective when patterning on the silicon-rich SiN layers.

In summary, low-cost, high-efficiency PERC solar cells can be fabricated using mechanical scribing, because this technique offers fast processing time, mass production capability, and simple processing with minimum cost. Using a mechanical scriber for patterning the front metal contact, MS-PERC devices can be fabricated without involving the costly photolithography process, thereby realizing significant reduction in fabrication cost, which is essential for commercial solar power modules.

3.4 Silicon Point-Contact Concentrator Solar Cells

Recent studies on silicon point-contact concentrator solar cells (SPCSC) have predicted conversion efficiency exceeding 37 percent in highly idealized solar cells [2]. A two-dimensional analysis on an SPCSC device has verified such high conversion efficiency using a line diffusion process. Note the fundamental recombination limit imposed by the Auger recombination in silicon under highly idealized conditions is valid, when the effects of diffusion gradients and resulting electric losses are neglected. In addition, optimization of the acceptance angle for the solar light absorption in the cell is essential to enhance the limits of photogeneration. Three-dimensional modeling is required to predict the performance of an SPCSC solar cell. This model can predict an ideal thermodynamic limit for the SPCSC solar cells as a function of incident power density and ambient temperature. This three-dimensional model has predicted an idealized thermodynamic limit of 37.8 percent for an SPCSC device under an incident solar intensity of 36 W/cm^2 at an ambient temperature of 330 K or 57°C or 134.6°F. The upper limit to the conversion efficiency for the point-contact solar cell is reduced to 32.3 percent due to incomplete absorption, diffusion gradients and transport losses. Note the overall conversion efficiency of the SPCSC is dependent on all critical parameters involved such as surface combination velocity, bulk lifetime, emitter saturation currents, and the Auger recombination in the carrier density range. It is important to point out that incorporation of polycrystalline diffused emitters in the fabrication of SPCSCs is the principal reason for the ultra-high efficiency of these devices.

3.4.1 Device Modeling Parameters

A three-dimensional model of the SPCSC is necessary to predict the performance parameters of the device. The device can be fabricated on a high resistivity silicon n-type float-zone (FZ) material with resistivity ranging from 100 to 400 ohms-cm. The emitter regions are highly doped. This model considers mostly low-doped emitter regions. Emitter recombinations are treated as boundary conditions, which are characterized by emitter saturation current (J_{oe}), surface recombination velocity, and bulk lifetime. The model assumes various carrier densities in different regions in the point-contact cell structure. The three regions in the device include a diffusion region in the silicon, a collection region, and a stagnant region, as demonstrated in Figure 3.3 (a). The geometry of the diffusions is on the back of the silicon wafer, as shown in Figure 3.3 (b). The three regions are designated as 1, 2, and 3 and the parameter w is the width of the solar cell. In region 1, the solution is considered to be one dimensional; in region 2, the carrier density is assumed constant; and region 3 is used as a collection region, where a three-dimensional solution is needed to solve the problem. However, perpendicular to the back surface in region 2 [Figure 3.3 (b)], the gradient in the carrier density falls rapidly, approaching to zero. In such device modeling, the modeled recombination components such as the bulk Auger recombination, emitter recombination, surface recombination, and bulk Shockley–Read–Hall recombination must be determined as a function of incident power density. It is important to mention that the recombination efficiency for all these components is strictly dependent on the solar incident power density. Furthermore, above incident power density of 20 W/cm² the recombination efficiency is dominated by the Auger recombination.

In the case of a typical SPCSC device, the cell thickness (W) is selected close to 120 μm operating under an incident power density of 36 W/cm² to achieve the maximum conversion efficiency close to 29 percent as shown in Figure 3.4. It is important to mention that the bounds on the carrier density drop occur in the lateral direction between the n⁺ and p⁺ contacts in region 2. Since the diffusion length is much greater than the distance between the contacts, these gradients are entirely insignificant in these devices operating at maximum power levels. Note the conversion efficiencies of SPCSC devices are dependent on device geometry and the incident solar power density. It is extremely critical that the top cell surface must be texturized and dimensional corrections should be made for the lumped series resistance in the

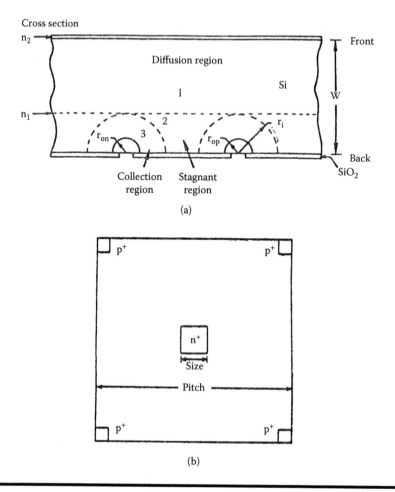

Figure 3.3 **Point-contact solar cell showing (a) the three regions involved and (b) diffusion geometry of the back surface of the silicon wafer.**

metallization on each cell to achieve maximum conversion efficiency. Conversion efficiency plots of these solar cells as a function of cell thickness (W), the diffusion size (S), and spacing (P) between the contacts are shown in Figure 3.4.

Modeling on the 120-μm thick SPCSC solar cell indicates that the quantum efficiency in the weakly doped regions remains in excess of 95 percent over the entire incident power density ranging from 1 to 36 W/cm^2 and under the air mass number of 1.5 (AM1.5). Note AM1 means the sun is directly overhead, that is, it is normal to the observer's head, whereas AM1.5 means that the path length is 1.5 times longer than that associated with AM1. On the other hand, a thinner device with thickness of 50 μm demonstrated quantum efficiency closer to 98 percent for incident power density as high as 56 W/cm^2 at AM1.5. This data excludes the effects from surface and emitter recombination components. However, according to the efficiency plots

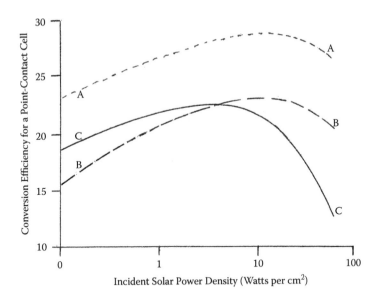

Device	Thickness (micron)	Diffusion (micron)	Spacing (micron)	Temperature (deg. C)
A	120	16	50	25
B	86	13	50	25
C	86	10	100	27

Figure 3.4 Conversion efficiency of a point-contact solar cell as a function of various parameters.

shown in Figure 3.4, higher conversion efficiency occurs only with thick devices. The higher the quantum efficiency, the higher the conversion efficiency will be. The effects of reduced quantum efficiencies and increased voltage drops are evident from the conversion efficiency versus incident power density curves shown in Figure 3.4.

3.4.2 Carrier Density in Various Regions of the Device

The carrier density changes in the boundary between the one- and three-dimensional regions shown in Figure 3.3 and are dependent on the dimensional parameters such as diffusion size (S) and spacing or the pitch (P) between the identical contacts shown in Figure 3.4. The density variation analysis is a function of radius, which increases linearly as the spacing between the contacts increases. This radius can be written as

$$r_1 = [P\,D_A/25.13]\,[1/D^2{}_p + 1/D^2{}_n] \qquad (3.1)$$

where P is the pitch or spacing between the identical contacts, D is the diffusion constant or coefficient, D_A is the ambipolar diffusion coefficient, and subscripts p and n stand for the p- and n-diffusion region, respectively.

This radius, when assumed as a hemispherical radius, will offer the maximum use of the symmetry of this particular modeling analysis. This hemispherical radius for the n+ diffusion is written as r_{on}. The three-dimensional drop in the carrier density from radius r_1 to radius r_{on} can be written as

$$[\Delta n_n] = [J(r_1)\ P^2/(12.57)q\ D_n]\ [1/r_{on} - 1/r_1] \tag{3.2}$$

where $J(r_1)$ is the current density in this region circled with radius r_1, q is the current charge, D_n is the diffusion coefficient for the n region, and Δn_n is the three-dimensional drop in the carrier density into the n-type contact as shown on the back of the silicon wafer (Figure 3.4). The expression for the drop into the p-type contact can be written as

$$[\Delta n_p] = [J(r_1)\ P^2/(12.57)q\ D_p][1/r_{op} - 1/r_1] \tag{3.3}$$

where D_p is the diffusion coefficient for the p region.

It is important to mention that a three-dimensional numerical solution is necessary to incorporate the effects of bulk generation and recombination. For a one-dimensional region, the total current is zero so the current density represents the flux of electrons or holes from the front to the back of this solar cell. Modeling analysis of this solar cell near the maximum power point becomes less sensitive to the effects of recombination in the base of the carrier energy profile, which makes it very easy to model the open-circuit voltage.

Note the carrier density is assumed to be continuous across the boundary between the region of uniform carrier density and the three-dimensional collection region, designated as regions 2 and 3 in Figure 3.3. Under this assumption, the current inside the three-dimensional region is assumed to be photogenerated current minus recombination current, already considered for regions 1 and 2. Higher average carrier densities are possible by reducing the emitter recombination with reduced emitter area. Increased current density occurs with steep gradients near the contacts. The drop in carrier density is inversely proportional to the diffusion size.

3.4.3 Terminal Voltage

Once the carrier density is determined, the terminal voltage can be estimated. This voltage can also be calculated from the carrier densities at two space-charge region edges with additional terms due to the voltage drops from the three-dimensional gradients into the contacts. The net terminal voltage (Vt) neglecting the carrier density voltage drops in the cell can be written as

$$V_t = [2kT/q] \; [\ln (n_{on} \, n_{op}/n_1 \, n_i)] \tag{3.4}$$

where the quantity $(2kT/q)$ is equal to 0.0258 at room temperature of 300 K, n_1 is the carrier density in region 1, and n_i the intrinsic carrier density. It is important to point out that steep gradients can result in non-negligible electric fields that decrease the terminal voltage magnitude. The ratio of the three-dimensional carrier density drop to the absolute carrier density drop determines the terminal voltage.

3.4.4 Photogeneration Profile of the Solar Cell

The photogeneration profile of this device consists of both a delta function of degeneration and a uniform generation in the one-dimensional region of the cell and its magnitude is proportional to the ratio of two diffusion coefficients. In other words, the correct average point of generation is determined by the integration of an AM1.5 spectrum as a function of depth in the silicon wafer. The most practical approach to find the average point of generation is to allow one pass of the photon spectrum through the solar cell thickness at the angle that would result from normal incidence onto the texturized surface of a <100> silicon wafer. The rest of the photogeneration can be assumed as uniform for the texturized wafers. The photogeneration for a 30-degree acceptance angle is considered optimum for many concentrator applications. The impact of photogeneration can be seen in the numerical integration in the one-dimensional region. Note the one-dimensional conversion efficiency for the 120-μm thick cell shown in Figure 3.4 is improved slightly from an efficiency of 27.66 percent to 27.75 percent just by maintaining a delta function of photogeneration at the front surface of the cell, which improved further to 27.94 percent by taking into account the texturization effect. The effect on the carrier density profile is small at the maximum power operating point. The one-dimensional model predicts higher carrier densities at the front of the cell (about 8 percent) and lower carrier densities at the back of the cell (about 6 percent at the p^+ depletion region edge). When the average carrier density in the solar cell increases as the emitter coverage fraction decreases, the recombination effects due to surface, bulk Shockley–Read–Hall, and Auger increase.

3.4.5 Techniques to Increase the Conversion and Quantum Efficiencies of the Cells

As stated earlier, higher concentration efficiency is possible with a 120-μm thick solar cell, which can be further enhanced by reducing the series resistance to half. This can be possible with a 4-μm aluminum metallization process. Using this process, the solar cell demonstrated conversion efficiency better than 28 percent in the incident power density range from 4 to 15 W/cm^2, which is equivalent to 40 to 150 suns. The conversion efficiency is still close to 25.6 percent at the highest incident power density exceeding 53 W/cm^2.

Emitter spacing, cell thickness, and diffusion size must be optimized for higher conversion efficiency. For a specific pitch or spacing optimum operating conditions can be satisfied for square diffusion sizes of 8 and 20 μm, which can yield higher quantum and conversion efficiencies. The current, a measure of quantum efficiency, is reduced monotonically as the diffusion size is reduced. This means higher quantum efficiency requires diffusion sizes exceeding 16 μm. An optimum coverage fraction exists at a spacing of 50 μm with an illumination density or incident power density of 10 W/cm². This means that the optimum coverage is for diffusions between 10 and 16 μm. It is important to mention that solar cells that do not have texturized front surfaces will have lower conversion and quantum efficiencies due to decreased photogeneration. The top curve shown in Figure 3.4 provides highest conversion efficiencies because of 16 by 16 μm diffusion sizes and a spacing of 50 μm. The quantum efficiency can be modeled by measurement of short-circuit current per incident watt at low incident power levels. Typical short-circuit current varies from 0.42 A per incident watt for a texturized cell to 0.35 A per incident watt for untexturized cells. As mentioned before, high carrier densities increase the Auger recombination, which will lead to poor quantum efficiency compared to devices with larger coverage fractions. In contrast to the current, the open-circuit voltages are always higher in cells with smaller coverage fractions because the three-dimensional gradients play a very small role in the absence of currents from the terminals.

3.4.6 Critical Design Parameter Requirements for Higher Solar Cell Performance

Significant performance improvement can be achieved in conversion and quantum efficiencies by optimizing certain dimensional parameters and by reducing the adverse effects of recombination. The Auger and the emitter recombination above the incident power densities of 20 W/cm² are the dominant parasitic losses as far as the conversion efficiency is concerned. The effect of the Auger recombination can be reduced, but the emitter recombination cannot be reduced further. Since the Auger coefficient is volume-dependent recombination, both the conversion and quantum efficiencies will improve as the cell becomes thinner up to a point where photogeneration is lost due to the weak absorption coefficient for the incident light that is near the bandgap in energy.

Optimum point-contact geometry must be determined for a specific cell thickness and responsivity. Contours of constant efficiency shown in Figure 3.5 can be developed as a function of diffusion spacing, diffusion size, and the relevant modeling parameters. At an incident power density of 10 W/cm² or 100 suns (10 suns is equal to 1 W/cm²) and at 300 K temperature, the projected conversion efficiency is about 30 percent as shown in Figure 3.5 (a). Contours of constant efficiency at the incident power density 36 W/cm² are shown in Figure 3.5 (b) and the projected efficiencies are between 30.5 and 31.0 percent, because the three-dimensional effects that oppose the deployment of small coverage fractions become more extreme at

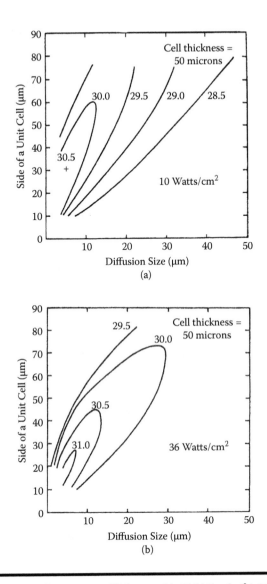

Figure 3.5 Contours of constant efficiency: (a) 10 W/cm²; (b) 36 W/cm².

high current densities. However, in these plots the optimum with respect to the diffusion size becomes much less sharp at higher incident power densities. These constant efficiency contours show fairly sharp optimum conversion values with respect to diffusion size for a given spacing. The cell efficiency is about 2 percent higher with optimum diffusion size than the limit as the diffusions approach full coverage of the back.

The most important design concept is to select the correct coverage fraction for each emitter type while using the smallest possible dimensions. This design criterion can be applied to other possible device geometries. Always consider the line diffusions as in interdigitated back contact solar cell, but with a spacing of undoped oxide-silicon interface between the n⁺ and p⁺ lines to achieve an optimum emitter coverage fraction. To achieve small emitter coverage fractions less than 10 percent, the diffusion lines must be spaced distances comparable to or greater than the cell thickness.

Efficiency of SPCSC is strictly dependent on the best emitter technology available for the point-contact devices rather than the diffused emitters using the current processes. Preliminary studies performed by the author on potential emitter technologies indicate that semi-insulating polysilicon with oxide-silicon spacing (SIPOS) offers the optimum performance because the emitter saturation current density of 1.2×10^{-14} W/cm² is possible for n-type emitters and 3.5×10^{-13} W/cm² for p-type emitters a room temperature of 300 K. If these current densities are applied to the 50-μm thick solar cells, conversion efficiencies better than 31.5 percent are possible using a pitch or spacing of 40 μm with 24-μm n-type contacts and 12-μm p-type contacts. Projected conversion efficiencies for optimized solar cells using SIPOS contacts and diffused emitters are shown in Figure 3.6. The optimized solar cell with diffused emitters has 12-μm square contacts for both n-type and p-type emitters. This particular device shows a conversion efficiency close to 30.6 percent at the design incident power density of 36 W/cm² at a room temperature of 300 K.

3.4.7 Conclusions on SPCSC Solar Cells

In summary, one can conclude that SPCSC solar cells can obtain conversion efficiencies better than 28 percent at incident power density of 15 W/cm² (i.e., at 150 suns) at 300 K. In the design of such cells, emitter recombination is the major limiting factor, particularly with thinner devices. Furthermore, the conversion efficiency can be further improved by 2 or 3 percent by reducing the cell thickness and optimizing the cell optics. If the lowest recombination emitters are incorporated in the SPCSC solar cell, conversion efficiencies better than 32 percent can be achieved.

3.5 V-Groove Multijunction (VGMJ) Solar Cells

The V-groove multijunction (VGMJ) solar cell consists of an array of many individual diode elements connected in series to generate a high-voltage low-current output This represents a fundamental difference between a conventional single-junction solar cell, in which the output voltage is less than one volt. All cell elements are formed simultaneously from a single silicon wafer using a V-groove etching process. Computer simulation predicts a conversion efficiency better than 28 percent for this device when operated in sunlight concentrated 100 times or

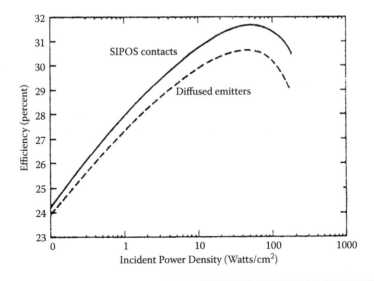

Figure 3.6 **Projected conversion efficiencies for a solar cell incorporating SIPOS and diffusion emitters.**

more (suns > 100). The major advantage of this solar cell over other conventional silicon cells include a conversion efficiency in excess of 25 percent after considering various losses, assuming modest bulk carrier lifetimes, a higher open-circuit voltage, a very low series resistance, a simple fabrication process involving one mask, and reliable and robust environmental protection provided by a glass front surface with improved mechanical integrity.

3.5.1 *Introduction*

It is important to mention that conventional silicon solar cells have not demonstrated conversion efficiencies exceeding 18 percent even under sunlight concentrated 50 times or more. On the other hand, VGMJ solar cells offer the potential for achieving conversion efficiencies in the range of 20 to 28 percent with sun concentration factor (SCF) from 10 to 10,000 or suns ranging from 10 to 10,000. In addition, the fabrication of the VGMJ solar cell requires only a single photomasking step and employs the most reliable semiconductor processing techniques, thereby realizing significant reduction in fabrication and mass production costs for the solar modules.

Studies performed by the author on photovoltaic (PV) devices reveal that concentration of sunlight can realize significant reduction in the cost of solar power generating system by substituting low-cost refractive and reflective materials for the expensive semiconductor materials. The studies further reveal that in a solar power system employing sunlight concentrated 50 or more times, the costs of the concentrator and support components may represent a major solar system cost. As a result of this, the solar power generation cost on a dollar/watt basis becomes very sensitive to the conversion efficiency of the PV elements. Therefore, the goal for achieving the highest conversion efficiency possible for the PV cells becomes a major economic decision for deployment of solar power generation systems. Note the development of high-efficiency silicon solar cells under concentrated sunlight conditions requires clear understanding of the physics that limit the PV device performance. Computer software programs are available for modeling both the optical and electrical performance capabilities of the silicon-based PV devices. A software program called PERSPEC [3] has been successfully used to design and optimize a silicon solar intensity sensor intended for use in sunlight concentrated up to 4000 times (i.e., suns = 4000). This computer program is best suited for the development and optimization of silicon-based VGMJ solar cells.

3.5.2 Description and Critical Elements of the VGMJ Solar Cell

Structural details of the VGMJ solar cell are shown in Figure 3.7. Note the solar light incident on the cell comes from above the glass cover and through the glass medium. The 7070 glass offers maximum transmission efficiency to the solar incident light and provides the environmental protection for the solar cell. Note this particular 7070 glass has a thermal expansion coefficient very close to that of silicon and its composition consists of low-alkali-potash-lithium-borosilicate. It is important to mention that glasses with low alkali content have not only lower losses, but also exhibit greater temperature stability up to 400°C.

This particular silicon VGMJ solar cell consists of several individual p⁺-i-n⁺ diode elements connected in series. The diodes use lightly doped p- or n-type silicon. The trapezoidal shape of the diode element as illustrated in Figure 3.7 is defined by the anisotropic etching (100) process for the silicon through a thermally grown silicon

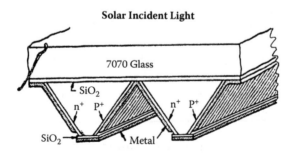

Figure 3.7 Isometric view of the V-groove multijunction (VGMJ) solar cell showing various metal sections, semiconductor layers, and 7070 glass cover, which is illuminated by incident light from the top.

diode layer. The silicon is supported by mounting it on Corning-7070 glass using an appropriate bonding technique. The diode elements can be made as long as permitted by the dimensions of the silicon wafer on which they are fabricated. Assuming a typical diode element width of 100 μm, about 100 diodes per centimeter of cell length can be fabricated. At 100 suns concentration factor, nominal cell electrical parameters such as current and voltage are roughly 33 mA/cm of the cell width and 70 V/cm of the cell length. Higher output voltage can be achieved by connecting the cell elements in series configurations. Conversely, higher output current can be achieved by connecting the diode elements in parallel configurations.

3.5.3 Fabrication Procedure for VGMJ Cells

The major steps of the fabrication procedure for the silicon-based VGMJ solar cell are shown in Figure 3.8. The fabrication of the VGMJ cell requires only one photomasking step, which is used to define the silicon dioxide V-groove matching

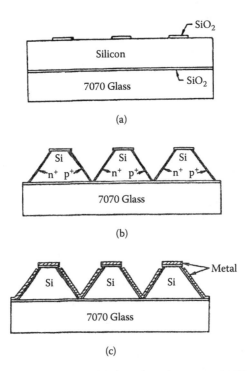

Figure 3.8 Fabrication steps for the VGMJ solar cell capable of yielding very high conversion efficiencies under various sunlight conditions. (a) Growth of the silicon dioxide layer, silicon oxide wafer to glass, and process for etching V-groove pattern windows in silicon. (b) V-groove etching down to 7070 glass, ion implant junction regions, and anneal implants. (c) The deposition of aluminum metal.

mask as illustrated in Figure 3.8 (a). Formation of the n⁺ and p⁺ junction regions [Figure 3.8 (b)] of the cell requires no masking steps, because it is accomplished by ion implanting the dopant atoms simultaneously. The ion implantations are followed by an annealing step to actuate the implanted dopant atoms and remove implantation damage. The deposited aluminum metal on the top of the oxide strips serves as a reflector, which enhances the light trapping capability of the cell as shown in Figure 3.8 (c).

3.5.4 Performance Parameters of VGMJ Cells

The most important performance parameter is the light-generated current in the VGMJ cell element, which can be computed from the one-dimensional model described in the PERSPEC computer simulation program. In the case of silicon solar cells, the excess carrier lifetime in the bulk region plays a critical role in determining the conversion efficiency through its rigid control of the collection efficiency, open-circuit voltage and fill factor. The effects of the carrier lifetime on these parameters in the VGMJ cells can be analyzed with the PERSPEC software by treating the device as a one-dimensional model. The optical generation of hole-electron pairs in the one-dimensional structure can be modeled by the uniform volume generation rate. The light-generated current (I_L) in the VGMJ element can be written as

$$I_L = [J_L\, F_{SC}\, (1 - R_0)\, \eta_{FC}\, A_E]$$ (3.5)

where J_L is the light-generated current density, which is equal to 40.5 mA/cm² in the silicon material produced by AM1 solar illumination, F_{SC} is the sunlight concentration factor, R_0 is the ratio of the number of photons that are reflected to the number of incident photons and has a typical value of 7 percent or 0.07, η_{FC} is the fundamental collection efficiency and has a typical value of 94 percent, and A_E is the light-illuminated area of the VGMJ element and is equal to the product of element length and element width. Computer generated plots of light-generated current (mA) as a function of sun concentration factor and light generated current density are shown in Figures 3.9 (a), 3.9 (b), 3.9 (c), and 3.9 (d) for sun concentration factors of 10, 100, 1000, and 10,000, respectively. These plots indicate that the light-generated current in the VGMJ elements increases with increase in sun concentration factor, which will lead to higher open-circuit voltages and conversion efficiencies of these solar cells. It is important to point out that most of the optically generated carriers in the VGMJ cell are produced below its illuminated surface, which is always closer to the collection junction.

3.5.4.1 Collection Efficiency of the VGMJ Solar Cell

One must distinguish between the total collection efficiency, fundamental collection efficiency, and internal collection efficiency in a solar cell. The total collection

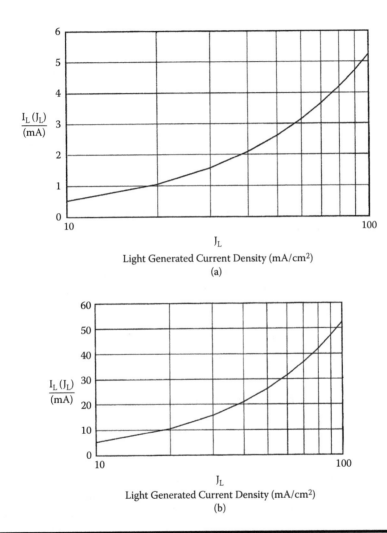

Figure 3.9   Light-generated current at a sun concentration factor of 10 (a), 100 (b) and 1000 (c).   *Continued*

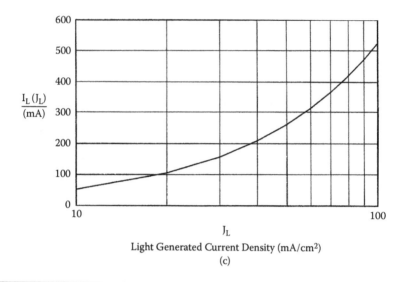

Figure 3.9 *Continued.*

efficiency is defined as the ratio of the number of hole-electron pairs collected by the cell surface to the number of photons with energy greater than or equal to the bandgap incident upon the top surface of the cell. Four distinct factors, the inactive cell area, the reflection loss, the fundamental collection efficiency, and the internal collection efficiency of the solar cell, control the total collection efficiency of the solar cell. The fundamental collection efficiency plays a major role in establishing the upper limit for the total collection efficiency.

3.5.4.2 Fundamental Collection Efficiency

The fundamental collection efficiency is dependent on the silicon layer thickness and the sunlight illuminations. Typical fundamental collection efficiency for a silicon cell as a function of silicon layer thickness under AM1 sunlight illumination is depicted in Figure 3.10. Note higher silicon thickness will increase the cost and the complexity of the cell. When a photon with energy level greater than or equal to the bandgap energy (E_g) enters a silicon wafer layer of finite thickness, it can either be absorbed to generate a hole-electron pair or to propagate through the silicon layer without undergoing a generative interaction, which defines the fundamental collection efficiency as mentioned above. It is evident from this figure that a silicon layer thickness of about 100 µm is required to achieve a fundamental collection efficiency of 90 percent under AM1 sunlight conditions.

The inactive area between the individual V-groove elements for electrical isolation must be kept to a minimum with proper control of the width of the openings in the V-groove etching masks. This particular area must not exceed 2 percent of

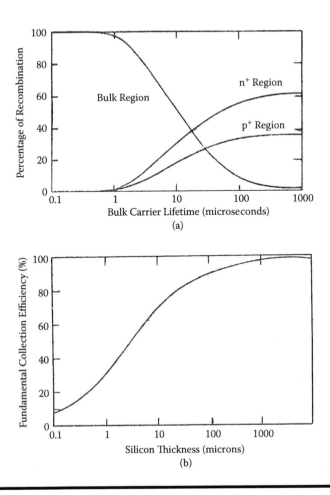

Figure 3.10 Performance parameters of a one-dimensional VGMJ solar cell showing (a) distribution of recombination and (b) fundamental collection efficiency.

the illuminated area of the VGMJ cell to maintain high total collection efficiency. It is interesting to mention that the inactive area loss in the VGMJ solar cell is similar to the shadowing loss observed in a conventional planar silicon solar cell. However, the shadowing loss in a conventional silicon solar cell is much larger than the inactive area loss, which is typically around 10 percent or higher under concentrated sunlight conditions.

3.5.4.3 Internal Collection Efficiency

In a VGMJ solar cell, photons can be reflected internally from the metal-covered regions on the back of the cell or from the metallized-junction regions. The internal collection efficiency of a cell is defined as the ratio of the number of hole-

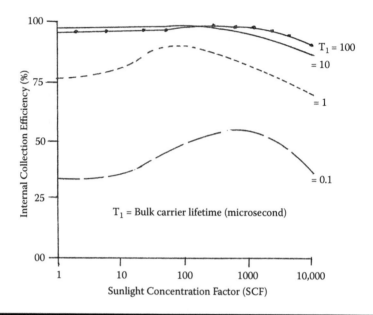

Figure 3.11   Internal collection efficiency of the one-dimensional VGMJ solar cell as a function of bulk lifetime (T_1) and sun concentration factor.

electron pairs that are collected by the cell to the number of photons generated within. The one-dimensional model of the VGMJ cell predicts that bulk carrier lifetimes greater than 100 μsec are needed to achieve an internal collection better than 95 percent with the sunlight concentration factor close to 1000:1 as shown in Figure 3.11. Note total recombination effects also affect the internal and total collection efficiencies. Distribution of recombination in a one-dimensional VGMJ cell as a function of bulk carrier lifetime is illustrated in Figure 3.11. The internal collection efficiency will boost the fundamental collection efficiency for the VGMJ cell, which will be greater than that of a conventional planar cell with thickness equal to the height of the trapezoidal elements. This particular thickness, called the "effective optical thickness" of the V-groove structure, is shown in Figure 3.7. An estimate for the effective optical thickness can be made by using ray tracing techniques. A VGMJ cell fabricated from a 50-μm thick silicon layer can be an effective optical thickness of over 290 μm, resulting in a fundamental collection efficiency exceeding 93 percent. However, a VGMJ solar cell using a thin silicon layer would require a long carrier lifetime to maintain high total collection efficiency. Note a 100-μm bulk carrier lifetime offers an increase in open-circuit voltage in the one-dimensional VGMJ cell structure over that 73 mV at 1 sun, 51 mV at 100 suns, 31 mV at 1000 suns, and 30 mV at 10,000 suns. The effects of sunlight concentration factors and bulk carrier lifetimes are displayed in Figure 3.11. The rise in internal collection efficiency from sunlight concentration factors is due to the field-enhancement effect, which is caused by the electric field in the bulk region of the cell. The subsequent roll off (Figure 3.11) in the internal collection efficiency at high sunlight concentration factors is caused partly by a decrease in the carrier diffusion length due to a reduction in the carrier mobilities through carrier-to-carrier scattering effects. The above statements indicate that one-dimensional VGMJ cell modeling must be undertaken involving a bulk carrier lifetime of 10 μsec and sunlight concentration factors from 1 to 1000 suns to obtain an internal collection efficiency of 95 percent or higher.

3.5.4.4 Reflection Loss in the VGMJ Cell

Reflection loss in a VGMJ solar cell is contributed by two distinct sources. In the fabrication of a VGMJ cell, the oxidized silicon wafer is bonded directly to the 7070 glass substrate as shown in Figure 3.7. In this case, more than 20 percent of the incident photons with the product $h\upsilon$ greater than the energy gap E_g in the AM1 solar spectrum are lost due to reflection. This reflection loss is quite large compared to roughly less than 5 percent observed in nonreflective silicon solar cells.

This high reflection loss in a VGMJ cell can be reduced through antireflection coatings. One coating is assumed to make direct contact with the silicon surface, while the second coating covers the illuminated surface of the 7070 glass. Deployment of a 550-Å tantalum oxide coating on the silicon and a 1550-Å coating of magnesium fluoride layer on the 7070 glass surface can reduce the reflection loss well below 6 percent. Although, the antireflection coatings do cut down the reflection losses, they also result in high surface recombination velocity at the illumination silicon surface. Techniques are available to reduce the surface recombination velocity. The internal collection efficiency is controlled primarily by bulk recombination, which can be modeled by the one-dimensional VGMJ cell model.

3.5.4.5 Open-Circuit Voltage and Voltage Factor

Higher open-circuit voltages are essential to achieve higher conversion efficiencies from solar cells. The ratio of the solar cell open-circuit voltage to the potential differences corresponding to the bandgap is called the voltage factor. In the case of p^+-i-n^+ silicon solar cells, the open circuit is mostly caused by the photogenerated carriers neutralizing the space charge near the junction regions of the cell. Note the concentration and spatial distribution of the optically generated carriers are determined by the nonlinear semiconductor carrier transport equations and Poisson equation. Computer models to determine the open-circuit voltage must take into account all the effects of doping-dependent lifetimes, Auger recombination, doping- and field-dependent mobilities, carrier-to-carrier scattering, and bandgap narrowing. The open-circuit voltage in such p-i-n solar cells is seriously limited by the reduction of the emitter efficiency of the junction regions at high excess carrier concentrations. A one-dimensional VGMJ cell model can predict the open-circuit voltage magnitudes as a function of sun concentration factors regardless of the length of the excess carrier lifetime in the bulk region. This one-dimensional model predicts the open-circuit voltage saturation at about 690 mV at 1 sun, 748 mV at 10 suns, 808 mV at 100 suns, 867 mV at 1000 suns and 920 mV at 10,000 suns. The emitter efficiency degradation causes the saturation of open-circuit voltage with increasing bulk carrier lifetime at a fixed concentration factor. Note for short bulk carrier lifetimes, most of the recombination occurs in the bulk region while for long carrier lifetimes, the majority of the recombination occurs at the junction regions as illustrated in Figure 3.11. Once the bulk excess carrier concentration saturates, the open-circuit voltage will no longer increase with increasing bulk lifetimes.

3.5.4.6 Fill Factor (FF) of a Cell

Fill factor is defined as the ratio of the maximum power output of a solar cell to the product of its open-circuit voltage and short-circuit current. The fill factor (FF) for a cell can be written as

$$\text{FF} = [I_{max} (mkT/q) \ln (I_{max} + I_{sc}/I_s)]/[V_{oc} I_{sc}] \tag{3.6}$$

where I_{max} is the maximum output current, m is a constant with typical value of unity, K is the Boltzmann's constant, T is the temperature (K), q is the unit charge, I_{sc} is the short-circuit current, I_s is the saturation current, and V_{oc} is the open-circuit voltage. The maximum current can be determined by the following equation:

$$[I_{max}/I_{max} + I_{sc}] = [\ln (I_s/I_{max} + I_s)] \tag{3.7}$$

Determining a precise value for the fill factor requires numerical solutions of the semiconductor carrier transport equations in conjunction with the Poisson equation. Two factors that affect the fill factor of a solar cell are the emitter efficiency degradation and the series resistance. At the maximum power point, the excess carrier concentration in the bulk region of the VGMJ cell is significantly reduced from its value under the open-circuit voltage condition. The fill factor is strictly dependent on the bulk carrier lifetimes and sunlight concentration factors ranging from 1 to 1000 suns. Typical values of FF with one-dimensional VGMJ cell of 0.8 or greater are possible from 1 to 1000 suns with bulk lifetimes of 50 µsec or greater. With a bulk carrier lifetime of 50 µsec the fill factor will be better than 0.8 at 300 suns.

Two components of the series resistance, namely, the internal resistance of the individual elements and the external series resistance due to cell metallization and the interconnect conductors, affect the open-circuit voltage. The ohmic power loss in the internal series resistance of the VGMJ cell is very low due to the small dimensions of its elements and the significant conductivity modulation of the bulk region in these elements, when the cell is operated under concentrated sunlight conditions ranging from 1 to 1000 suns. However, this loss at 10,000 suns is significant and can degrade the fill factor close to 0.74 for bulk lifetimes of 10 µsec or longer. Note it is extremely easy to keep the power loss in the external series resistance to below 1 percent even at sunlight concentration factor of 1000 suns. This would require an interconnect conductor cross-section of only about 0.02 mm² per centimeter of the running length. Similarly, the metallization to connect the adjacent elements on the wafer can be accomplished with minimum resistance. Metal runs with 1 mm thickness would yield less than 0.1 percent series resistance power loss at 1000 suns.

3.5.4.7 Total Conversion Efficiency of a VGMJ Solar Cell

Note the overall conversion efficiency or the total conversion efficiency of a silicon solar cell is dependent on all the parameters and effects as illustrated in Figure 3.12. A prototype and optimized version of the VGMJ cell illustrated in Figure 3.7 measuring 4900 μm on a side and with 43 series connected diode elements was fabricated following the fabrication steps shown in Figure 3.8 [3]. The optimized version of the device provided an open-circuit voltage of 0.702 V per diode and a short-circuit current of 44.3 mA. The optimized cell design configuration provided a fill factor of 0.8 and an open circuit of 30.2 V at 300 K and under 300 suns concentration conditions. Electrical performance characteristics and design parameters of the optimized version of the VGMJ silicon solar cell are summarized in Table 3.2.

$$\text{Total conversion efficiency (\%)} = [47\%] \ [0.98 \times .93 \times .94 \times .998 \times .75 \times .81]$$
$$= [47 \times .52] = [24.44] \ \%$$

It is important to point out that the spectral efficiency (η_{spec}) of silicon is strictly a function of the solar radiation wavelength, the depth of the p-n junction below the surface, the wafer thickness, the minority carrier lifetime and mobility in the base region, the absorption coefficient, quality of surface finish, purity and type of the silicon material, and the surface recombination velocity on the back surface of the device. It is extremely difficult to measure the spectral efficiency accurately under so many variables. However, this spectral efficiency can vary anywhere from 45 to 54 percent, according to various publications. Therefore, the total conversion efficiency can vary from 25 percent at a spectral efficiency of 48 percent to 26 percent at a spectral efficiency of 50 percent to 27 percent at a spectral efficiency of 52 percent and finally to 28 percent at a spectral efficiency of 54 percent. These predicted total conversion efficiencies for the VGMJ solar cells are the highest reported to date.

3.6 Potential Advantages of VGMJ Solar Cells

The following are the distinct advantages of VGMJ silicon solar cells:

- Conversion efficiencies exceed 25 percent using the existing technology.
- Internal collection efficiencies exceeding 95 percent are possible with only modest bulk carrier lifetimes.

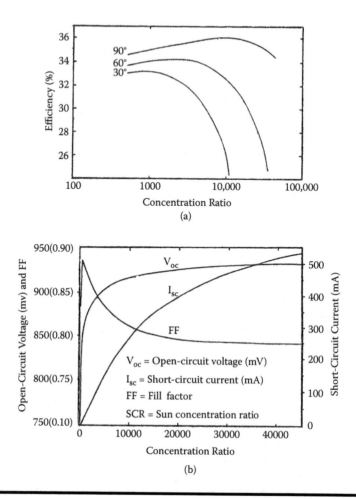

Figure 3.12 **Overall performance parameters for a silicon solar cell showing (a) efficiency and (b) other critical parameters as a function of sun concentration ratio.**

- Very low series resistance is necessary for efficient operation under sunlight condition of 1000 and beyond.
- Higher open-circuit voltages are possible with VGMJ cells than in planar solar cells of equivalent optical thickness.
- No current collection grid is required on the illuminated surface.
- This device provides excellent element-to-element uniformity.
- Reliable environmental protection is provided by the low-cost 7070 glass front surface.
- Simple and low-cost mask fabrication procedure offers maximum economy.
- Antireflection coatings provide reflection losses between 5 and 7 percent, which is not possible with other solar cell devices.

Table 3.2   Electrical Performance Characteristics and Design Parameters of an Optimized VGMJ Solar Cell

Performance Characteristics and Design Parameters	Values
Operating Conditions	
Sunlight concentration factor	300
Ambient temperature (K)	300
VGMJ Device Parameters	
Height (μm)	50
Bulk carrier lifetime (μsec)	50
Electrical Performance Parameters and Losses	
AM1 spectral efficiency of silicon (percent)	47
Collection efficiency due to inactive area (percent)	98
Reflection loss factor (percent)	93
Fundamental collection efficiency (percent)	94
Internal collection efficiency (percent)	99.8
Voltage factor	0.75
Fill factor (FF)	0.81

■ Device fabrication requires only one photomasking step, thereby providing significant savings in manufacturing costs.
■ Maximum design flexibility, ultra-high conversion efficiency, and lowest fabrication costs compared to other types of silicon solar cells are the principal benefits of VGMJ silicon solar cells; therefore, they are best suited for low-cost solar energy systems.

In summary, increasing the bulk carrier lifetime close to 50 μsec and using AR coatings to reduce reflection loss to less than 5 percent will further improve the conversion efficiency of the VGMJ cells. Studies performed by the author on these devices indicate that it is necessary to have a long bulk lifetime to achieve high conversion efficiency under sunlight concentration factors from 1 to 10,000. This requirement leads to higher open-circuit voltage with the incident illumination intensity. This means that the bulk carrier lifetime requirements for the 20 percent conversion efficiency at 1 sun is equivalent to roughly 229 μsec while at 100 suns it is about 5 μsec. The drop in internal collection efficiency at lower sunlight concentration factors as shown in Figure 3.7 occurs due to reduction in fill factor. A novel processing technique using two sequential ion implantations at an angle is available to form the junction regions of the trapezoidal elements in the cell. Bonding of silicon wafer of about 60-μm thickness to the 7070 glass substrate offers excellent and reliable mechanical support during processing. The VGMJ cell can offer conversion efficiencies close to 25 percent over a wide range of sun concentration factors. Thus, it allows a variety of concentrators, trackers, and heat removal techniques highly desirable to minimize the cost of solar energy systems. High design flexibility, ultra-high conversion efficiency, and lowest fabrication cost for the VGMJ solar cells offer a significant advantage over other types of silicon solar cells in designing an economically competitive solar energy system.

3.7 Multiple-Quantum-Well (MQW) GaAs Solar Cells

The conversion efficiency of a single p-i-n junction solar cell can be enhanced by incorporating quantum wells in the intrinsic region of the solar cell device. The incorporation of the multiple quantum wells has two distinct effects: the short-circuit current is increased because of the additional absorption of low-energy photons in the lower bandgap quantum wells and the open-circuit voltage is slightly decreased due to increased recombination of carriers trapped in the quantum well [4].

3.7.1 Introduction

The rates of escape, capture, and recombination of photoexcited carriers in quantum wells embedded in the intrinsic region of p-i-n devices such as GaAs-AlGaAs quantum wells play a key role in achieving optimum performance. This means that investigation of the performance of quantum well solar cells is strictly dependent on the dynamics of capture, escape, absorption, and recombination of the carriers in the quantum well regions. The additional photocurrent generated from the extension of the absorption spectrum to lower energy levels can cause a drop in the open-circuit voltage, which will limit the enhancement in the conversion efficiency due to incorporation of quantum wells. A balanced theory developed by various scientists in 1990 to 1991 indicates that the conversion efficiency of the quantum-well solar cell would not exceed the efficiency of the baseline bulk solar cell with optimum bandgap in case of radiative recombination dominance. These scientists further predicted that enhancement in both the open-circuit voltage and conversion efficiency is possible, if the device deploys the structure of multiple-quantum-wells (MQWs). Photoresponse calculations reveal that the insertion of quantum-wells in a solar cell could lead to improvement in photocurrent levels without much degradation in open-circuit voltage, provided the device employs MQWs with depths not exceeding 200 meV. Furthermore, material scientists are currently investigating potential quantum-well materials capable of providing the needed improvements in the conversion efficiencies of the MQW-based solar cells. However, the thermodynamic arguments predict that the ideal MQW solar cell device could not exceed the efficiency of an ideal ordinary solar cell using the same semiconductor materials, unless the carriers in the well are pumped by the absorption of a secondary photon.

3.7.2 Impact of Capture and Escape Times on Device Performance

The baseline bulk $Al_x\,Ga_{1-x}$ As/GaAs device provides a classic example of a p-i-n MQW solar cell. It is important to point out that quantum-well width and aluminum mole fraction (x) play critical roles in shaping the performance level of MQW solar cells. Published data on electron capture time as a function of well width indicate that the capture times vary from 0.5 psec at well width of 40 Å to 1.5 psec at well width of 95 Å to 1.0 psec at well width of 120 Å and back to 1.5 psec at well width of 180 Å with a mole fraction of 0.1 and under constant electric field of 10 kV/cm. The capture times versus well width exhibit oscillatory trends, which are due to the inclusion of additional eigenenergies with increasing well width. This indicates that every time an additional eigenenergy becomes bounded, the capture time decreases.

The electron escape time is strictly dependent on electric field for various well widths and aluminum mole fraction. The increase in electric field results in shorter escape times for the electrons. Further small mole fraction of aluminum yields faster escape time due to the decreased barrier height. Conversely, the electrons experience longer escape times in wider wells. Note electric field not only changes the position of the eigenenergy in the well, but also lowers the height of the "downhill" barrier due to tilting of the bands.

Similar trends have been observed for hole escape times. However, the holes generally escape faster than the electrons for a given set of parameters, which is due to the lower barrier heights of the quantum wells in the valence band. The effects of various parameters on the electron escape and hole escape times are quite evident from the values summarized in Table 3.3 and Table 3.4, respectively.

It is evident from the data that smaller mole fractions of aluminum have faster escape times due to the decrease in barrier height. The electrons have longer escape times in wider quantum-wells. It is also evident from the data that the holes escape faster than the electrons for the same parametric values, which is strictly due to the lower barrier heights of the quantum-wells in the valence band.

3.7.3 Performance Parameters for the Baseline Bulk $Al_x\,Ga_{1-x}$/GaAs Solar Cells

Before investigating the performance capabilities of the p-i-n MQW solar cells, performance parameters for the baseline bulk device must be established for comparison purposes. A top p-region thickness of 200 nm for the baseline device is considered

Table 3.3 Electron Escape Times as a Function of Electric Field, Aluminum Mole Fraction (x) and Quantum-Well Width (Picoseconds)

		Well Width (Å) and x = 0.1/0.3		
Parameter		50	100	150
Electric field (kV/cm)	10	0.8/22	1.7/130	2.2/235
	20	0.7/22	1.2/115	1.4/165
	30	0.7/22	0.9/100	0.9/137
	40	0.6/100	0.7/115	0.7/120

Table 3.4 Hole Escape Times as a Function of Electric Field, Aluminum Mole Fraction (x) and Quantum-Well Width (Picoseconds)

		Well Width (Å) and x = 0.1/0.3		
Parameter		50	100	150
Electric field (kV/cm)	10	2.0/25	3.5/58	3.7/85
	20	2.0/23	2.6/47	2.7/50
	30	1.9/22	1.8/33	1.7/35
	40	1.7/20	1.1/19	1.2/19

thick enough to provide the required electric field level and thin enough for most photogenerated carriers to be in the space charge region. Similarly, a 600-nm thick n region is able to provide enough absorption without excessive recombination. Even when the n-region thickness is increased to 2 μm, there is no evidence of improved efficiency. Typical doping levels of $9 \times 10^{17}/cm^3$ in the p region and $2.5 \times 10^{17}/cm^3$ in the n region are considered adequate for good overall device performance. The conversion efficiency for an $Al_x Ga_{1-x}$ GaAs device with no quantum wells and when x is equal to zero is about 17.4 percent [4] compared to an efficiency of 17.1 percent for the baseline solar cell when the mole fraction x is equal to 0.1. Note higher aluminum content creates a large bandgap, which increases the open-circuit voltage, but decreases the short-circuit current densities due to lack of absorption capability. This is quite evident from the data summarized in Table 3.5.

These values indicate that the open-circuit voltage increases linearly with the mole fraction of the aluminum, while the current density decrease is much steeper. As a result, the baseline devices are less efficient than the conventional GaAs p-n

Table 3.5   Open-Circuit Voltage and Short-Current Density of the Baseline Device with No Quantum Wells in the Intrinsic Region as a Function of Aluminum Mole Fraction (x)

Mole Fraction (x)	Short-Circuit Current Density, mA/cm²	Open-Circuit Voltage, V
0	23.4	0.65
0.1	20.1	0.71
0.2	16.2	0.80
0.3	13.2	0.92
0.4	10.5	1.14

junction solar cells. However, when quantum-wells are incorporated in $Al_x GA_{1-x}$ As Ga As devices, some efficiency enhancements have been observed, but these efficiency enhancements cannot beat the efficiency of the bulk GaAs device. Note that the decrease in efficiency for the baseline Al Ga As/GaAs is very small, thus making the aluminum ally the best candidate for possible efficiency enhancement when quantum-wells are incorporated in the MQW solar cells.

It is important to mention that the efficiency of 17.4 percent for the baseline GaAs solar cell is significantly lower than the optimum efficiency of 23.1 percent available from the bulk GaAs device. However, one can achieve both the higher open-circuit voltage and the conversion efficiency by using a higher quality epitaxial layer, even though the trap density in AlGaAs alloys is much higher than the trap density found in good quality GaAs material.

3.7.4 Electric Field Profiles and Carrier Density Distribution in AlGaAs Devices

The electric field profiles are dependent on the distance of the quantum-well from the surface in the Al GAAs/GaAs device, device output voltage, and the aluminum mole fraction. Relatively high electric fields exist at the interface between the doped regions and the intrinsic region. This results from the depletion of the carriers just inside the doped regions, where there is a steep gradient in the electric field profiles. Throughout the intrinsic region, the electric field remains constant at about 10 kV/cm when the device operates at its maximum power level of 0.79 V. In MQW solar cells, the doping is discrete, which can lead to local variations in the electric field. However, the doping in the intrinsic region has a finite value that affects the electric field magnitude in the quantum-well regions. Note as long as the intrinsic region doping level is below $10^{15}/cm^3$, the electric field would be large enough to collect carriers for the quantum-wells.

As far as the carrier density distribution is concerned, the electric field causes a sharp decrease in carrier concentration at the interface between the doped and intrinsic regions. Since the solar cells operate at a low forward bias level, the majority carriers diffuse into the intrinsic region giving a dark current in the forward direction. However, the solar cell current flows in the opposite direction. Thus, the minority electrons diffuse from the p region toward the space charge region.

Regarding the recombination rates in the quantum-well regions, the recombination rates of the bulk and MQW devices match very closely, except in the quantum-well regions, where recombination is significantly enhanced. Note as more carriers flow into the intrinsic region under increased bias conditions, the recombination rate increases. The recombination rate barely shifts in the doped regions as a function of voltage, while it increases significantly in the intrinsic regions within the wells. The current density consists of the diffusion and drift currents. Note under short-circuit conditions, the total current density is roughly 17.4 mA/cm², where its value is about 16 mA/cm².

3.7.5 Impact of Physical Dimensions of the Quantum-Well on Solar Cell Performance

The width of the quantum-well and the number of wells play a key role in shaping the performance of the MQW devices. It is important to note that thinner quantum-wells allow to be incorporated in the device, while the wider wells have fewer $AI_xGa_{1-x}As/GaAs$ interfaces and hence, fewer interface recombination rates. The wider the well, the less likely the carrier will be able to escape, which increases the likelihood of recombination in the well. The short-circuit current density increases with the increase in well width for a given number of wells. Note the wider wells have lower open-circuit voltages because of increased recombination due to the longer escape times in the wider wells. Losses in fill factors are associated with the increase in well recombination. Higher fill factors are possible with wider wells. A well width of 150 Å has a typical fill factor of about 0.71.

For a given number of MQWs, the conversion efficiency of the solar cell decreases as the aluminum mole fraction increases. However, the conversion efficiency improves with the increase in number of quantum-wells. Projected values of conversion efficiency as a function of number of quantum-wells and quantum-well width are summarized in Table 3.6.

It is evident from the data that the increase in conversion efficiency is extremely small as a function of number of wells for a given width of the quantum-well. The data further indicates that the efficiency decreases with increase in the aluminum mole fraction. It is quite evident that the efficiency suffers significantly with aluminum fractions of 0.3 and 0.4 due to high recombination rates. In other words, the recombination losses are the principal reason for degradation in the quantum-well solar cell performance. The open-circuit voltage suffers seriously from the increased recombination at the interface traps. In addition, the short-circuit current density begins to suffer when the trap density exceeds $5 \times 10^{11}/cm^3$.

Table 3.6 Conversion Efficiency of MQW Solar Cells as a Function of Well Width and Number of Wells (percent)

Well Width (Å)	Aluminum Mole Fraction (x)				No. of Wells		
	0.1	0.2	0.3	0.4	10	20	30
50	17.1	14.7	12.8	9.7	16.98	16.95	16.97
100	17.3	15.3	12.4	9.3	—	—	—
150	17.4	15.6	12.8	10.2	17.17	17.34	17.57

3.8 Summary

In this chapter solar cells capable of providing very high conversion efficiencies with minimum fabrication costs are described, with emphasis on simple design with no compromise in device reliability. A new method for patterning the rear passivation layers using a mechanical scriber is described, which is best suited for fabrication of high-efficiency passivated emitter and rear cell (PERC) solar cells. A conversion efficiency of better than 19.5 percent has been demonstrated for the mechanical scribe-based PERC solar cells. The mechanical scriber process shows a great potential for commercial applications by achieving conversion efficiencies exceeding 22 percent, while reducing significantly the fabrication costs with an expensive photolithography process. Three distinct types of rear contact patterns are discussed with significant reduction in contact area and contact resistance. Design criteria and performance parameters are defined for silicon point-contact concentrator solar cells. The three-dimensional case of a back-surface point-contact solar cell is discussed with emphasis on recombination components and the carrier density gradients on the geometrical design parameters. These solar cells using the diffused emitters have demonstrated conversion efficiencies better than 28 percent and 31 percent at incident solar power density of 15 W/cm^2 and 36 W/l^2, respectively. Incorporation of polycrystalline emitters could lead to further improvement in conversion efficiencies approaching to 32 percent. The conversion efficiency of the silicon point-contact concentrator is strictly dependent on incident power density, cell thickness, diffusion size, spacing or pitch, and ambient temperature. In addition, the combination components that affect the conversion efficiency are dependent on the incident power density. The Auger and emitter recombination are the most dominant parasitic losses and, therefore, these losses must be kept to a minimum, if higher conversion efficiency is the principal design requirement. This offers the highest efficiency with minimum production costs, thereby yielding the most cost-effective solar energy systems for commercial applications. V-groove multijunction (VGMJ) solar cells consist of an array of many individual diodes elements connected in series to generate high-voltage low-current output best suited for commercial solar-based power sources. Note all the elements can be formed simultaneously from a single wafer using a V-groove etching process and a simple one-mask fabrication process, leading to lowest production costs. High internal collection efficiency better than 98 percent, fundamental collection efficiency in excess of 94 percent, and reflection loss using antireflection coatings less than 5 percent offer high conversion efficiency for the VGMJ solar cell. Typical conversion efficiency for a VGMJ solar cell with only modest bulk carrier lifetimes is in excess of 24 percent when the device is operated in sunlight concentration factor of 100 (suns = 100). Output voltage as high as 70 V per centimeter of cell length is possible, which requires only a cell length of 1.63 cm to provide a conventional supply voltage of 110 volts. Excellent environmental protection and high reliability are provided by a front glass cover with high mechanical integrity. A simple one-mask fabrication process, minimum production costs, and higher conversion efficiency using only the existing technology are essential for economically competitive solar energy systems. The conversion efficiency of a simple p-i-n solar cell can be improved by incorporating quantum-wells in the intrinsic region of the device. However, the incorporation of the multiple-quantum-wells (MQWs) will increase short-circuit current due to absorption of low-energy photons in the lower bandgap quantum-wells and will decrease slightly the open-circuit voltage because of increased recombination of the carriers trapped in the quantum-wells. Although the increased recombination reduces the open-circuit voltage, limited enhancement in the conversion efficiency is possible by incorporating the quantum-wells in the intrinsic region. The conversion efficiency of 17.40 percent for the baseline $Al_x Ga_{1-x} As/GaAs$ device with no quantum-well is significantly lower than the optimum efficiency of GaAs p-n junction solar cells currently available. Note the conversion efficiency for a bulk GaAs solar cell is in excess of 24 percent even after taking into accounts all losses. It is important to mention that the inclusion of quantum-wells of optimum widths and using an appropriate number of quantum-wells can even improve the conversion efficiency of a bulk AlGaAS solar cell.

References

1. J.S. Moon, D.S. Kim, et al. "New method for patterning the real passivation layers of high-efficiency solar cell," *Journal of Materials Science: Materials in Electronics* 12 (2001): 605–607.
2. R.A. Sinton and R.M. Swanson. "Design criteria for silicon point-contact concentrated solar cells," *IEEE Transactions on Electron Devices* ED-34, No. 10 (1987): 2116–2122.
3. Terry I. Chappell. "The V-groove multi-junction solar cell," *IEEE Transactions on Electron Devices* ED-26, No. 7 (1979): 1091–1097.
4. S.M. Ramey and R. Khole. "Modeling of multiple-quantum well solar cells including capture, escape, and recombination of photoexcited carriers in quantum-wells," *IEEE Transactions on Electron Devices* 50, No. 5 (2003): 1181–1187.

Chapter 4

Techniques to Enhance Conversion Efficiencies of Solar Cells

4.1 Introduction

This chapter focuses on potential techniques for enhancing the conversion efficiencies of solar cells regardless of the material used in the fabrication of the devices. Intensity enhancement is the critical requirement to increase the conversion efficiency of a solar cell. Other methods to enhance cell conversion efficiency such as contact design configurations and materials, front cover surface with optimized performance as a function of the refractive index, antireflection (AR) coatings, oxidation process requirements for the silicon surface, light trapping techniques, doped region thickness to avoid significant contribution to recombination, grain size of the solar cell material, optical tilt angle for large solar arrays, materials with appropriate values of minority carrier lifetimes and surface recombination velocity, improved spectral response, surface layer thickness selection for high internal collection efficiency and total collection efficiency, bifacial modules and hemispherical mirrors, solar illumination intensity or sunlight concentration factors, use of nanotechnology materials (nanowires and nanocrystals), all-dielectric micro-concentrators (ADMCs), and two-axis solar trackers are discussed in detail, with emphasis on efficiency enhancement, cost, and reliability of the solar cells and modules. Each of these techniques or concepts will be described in terms of how it improves the conversion efficiency of the solar cell. The overall conversion efficiency of a solar cell can be expressed as

$$\eta_{o-conv} = [\text{spectral efficiency}][\text{collection efficiency}][\text{voltage factor}][\text{fill factor}]$$
$$= [\eta_{spec}][\eta_{coll}][V_{fact}] [FF] \tag{4.1}$$

where η stands for efficiency, the subscript coll stands for collector and FF stands for fill factor.

The total collection efficiency of the solar cell consists of four distinct components, namely, the fundamental collection efficiency, reflection efficiency or reflection loss factor, inactive area, and internal collection efficiency and its expression can be written as

$$\eta_{coll.} = [\text{inactive area (percent)}][\text{reflection efficiency}][\text{fundamental collection efficiency}][\text{internal collection efficiency}] \tag{4.2}$$

The impact of bulk carrier lifetime and sunlight concentration factor on internal collection efficiency will be discussed under solar illumination intensity or concentration factors. The overall conversion efficiency is dependent on several factors, such as interaction area of the solar cell device (typically 98 percent), reflection efficiency (typically 95 percent), fundamental collection efficiency (typically 95 percent), contact efficiency (typically 96 percent), internal collection efficiency (typically 98 percent), voltage factor (typically 0.9), fill factor (typically 0.75), and bulk carrier lifetime (ranging from 230 µsec at 1 sun to 5 µsec at 100 suns, where sun represents the sunlight concentration factor). The total collection efficiency, also known as the fundamental collection efficiency, is the major contributing source, which is a function of silicon cell thickness and the sunlight illumination factor. Typical values of fundamental collection efficiency for a one-dimensional silicon solar cell under AM1 sunlight illumination are summarized in Table 4.1.

The air mass (AM) term is frequently used in specifying the conversion efficiency or fundamental efficiency of a solar cell. AM refers to the intensity and the spectral distribution resulting from a given path length of sunlight through the atmosphere. AM1 means that the sun is directly overhead, that is, the sun's radiation is normal to the solar cell surface. AM1.5 indicates a sun position such that the path length is 1.5 times more than AM1.

4.2 Impact of Contact Performance and Design Parameters on Conversion Efficiency

Cell contacts and interconnect lines are needed to connect several solar devices in series and parallel configurations to meet specific voltage and current requirements of a solar energy power source. Contact resistance losses occur at the interface between the silicon solar cell and the metal contact, which can affect the cell performance. In

Table 4.1 Typical Values of Fundamental Collection Efficiencies for One-Dimensional Silicon Solar Cells with AM1 Sunlight Illumination

Silicon Thickness (μm)	Fundamental Collection Efficiency (%)
0.5	18
1.0	30
5	50
10	71
50	86
100	93

addition, contact resistance has a significant effect on the current-voltage characteristics of the cell. The main impact of the increased series resistance is to reduce the fill factor (FF), which is a parameter of the overall conversion efficiency [Equation (4.1)]. As far as the metal-semiconductor contact performance degradation is concerned, screen-printed front contacts using silver paste are degraded more smoothly compared to thermally vacuum-evaporated front contacts of titanium/palladium/silver (Ti/Pd/Ag). Front radical contacts are fabricated of Ti (0.5 μm)/Pd (0.5 μm)/Ag (3 μm) thermally vacuum evaporated and a tin oxide (SnO$_2$) layer of 80 nm thickness is deposited as antireflection coating [1]. The ohmic contacts are fabricated using a screen-printed technique involving deposition of silver/aluminum paste for back contact surface, deposition of silver paste for front contact surface, and deposition of aluminum paste on the back contact. Thermal cycles cause degradation in the shunt and series resistance components because of the thermal conductivity difference between the metallic contacts and silicon semiconductor. The loss of metal-semiconductor adherence by current pulses increases contact resistances to the front and back as well as surface leakages along the cell edges by the diffusion spikes. Studies performed by the author on various types indicate that screen-printed contacts are degraded more smoothly than thermally vacuum-evaporated front contacts of Ti/Pd/Ag. The studies further indicate that contact degradation plays a key role in reduction of conversion efficiency, but it does not degrade the electrical characteristics of the junction.

4.3 Intensity Enhancement in "Textured Optical Sheets" (TOS) Used in Solar Cells

Australian research scientists have proposed that by randomizing texturing of the surface of a sheet, the internal light intensity can be enhanced by a factor of 8n rather than 2n^2, where n is the refractive index of the sheet material. This argument

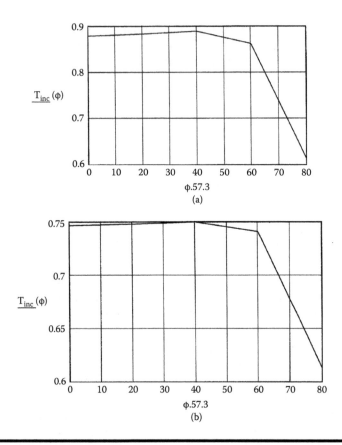

Figure 4.1 **Transmittivity of the incident radiation from air to flat dielectric surface as a function of incident angles when *n* = 2 (a), 3 (b), 4 (c), and 5 (d).**

Continued

is based on an analytical expression for the transmission factor for the light incident on a flat surface at an angle φ to the normal to the material with a refractive index *n*. The geometrical optics theory can be applied using Fermat's principal, where the family of rays is everywhere orthogonal to the wavefronts in an isotropic medium. The ray paths are straight lines in a homogeneous medium, but they can change directions at an interface between two different media according to Snell's laws of reflection and refraction, which can be deduced from Fermat's principle [2]. The electrical transmission factor for the solar cell (T_{esc}) when averaged over the surface wave and incident power wave within the medium can be written as

$$T_{esc} = (1/2n)[\{2n \cos\theta/n \cos\theta + \cos\varphi \}^2 + \{2n \cos\theta/\cos\varphi + n \cos\theta \}^2] \quad (4.3)$$

where *n* is the refractive index of the medium, φ is the angle of incidence, and θ is the external angle defined by Snell's law. But the energy in each medium is

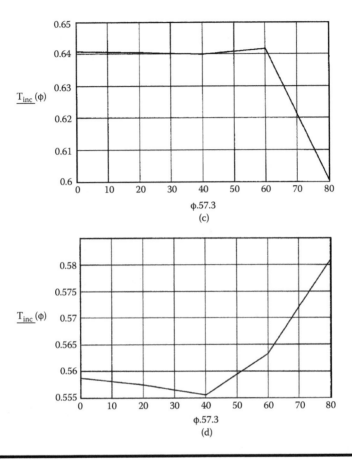

Figure 4.1 *Continued.*

represented by the Poynting vector. In essence, the Poynting vector represents the energy per second that crosses a unit area normal to the directions of the electric and magnetic vectors (E and H), that is, in the direction of wave propagation. This means that the transmittance is defined as the ratio of Poynting vectors associated with the incident and transmitted waves and this ratio is defined as $\cos\varphi/\cos\theta$. This means that the transmission factor for the incident wave can be written as

$$T_{inc}(\varphi) = (1/2n)(\cos\varphi/\cos\theta)[\{2n\cos\theta/n\cos\theta + \cos\varphi\}^2 + \{2n\cos\theta/\cos\theta + n\cos\varphi\}^2] \tag{4.4}$$

Transmittivity plots of incident radiation from air to flat surface as a function of incidence angle and refractive index are shown in Figures 4.1 (a) to (d) for given values of theta.

It is important to mention that an incident beam of light transmitted across a flat surface will be confined to a different area after the transmission to the change of

Table 4.2 Hemispherically Weighted Average Values of the Transmission Factor T_{esc} as a Function of Refractive Index n

Refractive Index n	Average Value of T_{esc}
2	0.84
3	0.72
4	0.64
5	0.55

angle at the interface. Furthermore, this form of expression as defined by Equation (4.4) has the advantage of always being less than unity, as illustrated by the plots shown in Figure 4.1, and of satisfying the energy conservation laws when combined with the corresponding expressions for the reflectance at the interface. Due to time-reversal variance, the calculated value of $T_{inc}(\varphi)$ transmittance into the material as a function of incident angle φ is equal to the value of transmittance parameter $T_{inc}(\theta)$. Now the equation relating the internal light intensity (I_{int}) to the incident light intensity can be written as

$$I_{int} = I_{inc}\,[(2n)^2][T_{inc}(\varphi)/T_{esc}] \tag{4.5}$$

When Equations (4.3) and (4.4) are inserted in Equation (4.5), it becomes quite evident that the enhancement is by a factor of $2n^2$. Hemispherically weighted average values of the transmission factor T_{esc} as a function of refractive index n are shown in Table 4.2.

4.4 Nanoparticle Plasmons Best Suited for Solar Absorption Enhancement

As mentioned earlier, efficiency enhancement and reduction in fabrication costs can be accomplished using various techniques. For example, thin-film technology, laser-based processing, and nanoparticle plasmons-based solar absorption offer higher conversion efficiencies, lower fabrication costs, and higher production yield desirable for solar panel production.

4.4.1 Nanotechnology Concepts to Enhance Solar Cell Conversion Efficiency

The conversion efficiency of the solar cell can be improved by enhancing the solar absorption capability of the surface of the device exposed to solar radiation.

Nanotechnology scientists believe that surface plasmon resonances of gold nano-particles can augment the conversion of photons into usable energy, thereby real-izing some improvement in conversion efficiency in the solar cells. An array made up of identical elliptical gold nanodisks with height of about 20 nm and minor and major axes of 40 and 120 nm, respectively, deposited on top of a thin tin oxide semiconductor film sensitized with dye molecules, has demonstrated improved solar absorption capability. Nanoparticles of identical size offer well-defined plasmonic response, which can be controlled by the polarization of the light. This led to further investigation into how much solar absorption is possible using this technology.

These scientists have demonstrated that sprinkling nanoparticles over the active face of the solar cells will enhance the performance of the solar cells [3]. Research scientists are currently investigating how many of these tiny particles, what sizes, what materials, and what geometrical configurations are needed to achieve opti-mum performance. Even small amounts of nanoparticles distributed over the light-collecting face of a photodiode or a solar cell could significantly enhance or suppress the photocurrent generated in the device. These nanoparticles could be solid silica nanospheres (typical radius = 60 nm) or solid gold nanospheres (typi-cal radius = 25 nm) or gold and silicon nanoshells with radii ranging from 62 nm to 116 nm. According to research scientists, the images when measured using an optical microscope with laser wavelengths of 532, 633, 785, and 980 nm exhibit bright spots with increased photocurrents and darker spots at reduced photocur-rents. The researchers further indicate that the silica nanospheres exhibit a con-sistent and uniform enhancement of the photocurrent over the laser wavelengths mentioned above, which is attributed to the nonresonant scattering properties of the nanoparticles. Based on the findings by researchers, a maximum photocur-rent enhancement of about 20 percent is possible with larger nanoshells at 980-nm laser excitation. The strongest photocurrent suppression close to 30 percent has been observed with 116-nm nanoshells at a laser wavelength of 633 nm. The research data indicate that nanoparticles with suitable dimensions could be used for enhancement of silicon-based solar cells at long wavelengths, provided the loss of photocurrent can be minimized at shorter wavelengths.

Solar cell scientists believe solar photonic technology can trap incident solar light on solar cell surfaces constructed with silicon nanowires of appropriate physical dimensions, which will lead to enhancement in the conversion efficiency of the cell.

4.5 Laser-Based Processing to Boost Conversion Efficiency and Reduce Production Costs for Solar Cells

Manufacturing costs, fabrication processing steps, optimization of production tools, and product standardization are necessary to obtain higher conversion efficiency and lower device cost simultaneously without compromise in device reliability. It is quite possible that different manufacturing technologies and the technological background of the equipment suppliers must be changed in the next 5 to 10 years to the stated cost-effective goal, if 40 percent or more compound annual growth rates are to be sustained. It is appears that there are enormous opportunities for laser-based systems to move from a competitive processing equipment type to a dominant one for critical processing requirements. Furthermore, continued worldwide solar power as a renewable energy type is critically dependent on several factors, such as economic drivers, environmental change requiring low carbon footprint production equipment, and technical enhancements in terms of higher conversion efficiency, lower production cost, and higher device yield. Implementation of changes in these areas can be satisfied through rapid deployment of laser technology and selecting appropriate laser wavelengths for optimum device performance [4].

4.5.1 Crystalline-Silicon Solar Cells Most Likely to Get Most Benefits from the Deployment of Laser Technology

Research scientists believe that crystalline silicon solar cells are most likely to derive the most benefits from the deployment of laser technology in manufacturing of solar cells. A current market survey on solar cells published in the *Photonic Spectra* dated April 2008 indicates that approximately 90 percent of all solar panels produced today are made from crystalline-silicon (c-Is) material and predicts that it will remain at 75 to 80 percent into 2015. Studies performed by the author on solar cell processing techniques indicate that there are a number of processing steps that can be accomplished with minimum cost, high reliability, and maximum yield with no compromise in accuracy or quality control. It is important to mention that thin-film solar cell manufacturing requires lasers for thin-film patterning, in which accuracy, consistency, and quality control are of critical importance. It is interesting to mention that the typical conversion efficiency of a c-Is solar cell in the year 2007 was in the range of 14 to 18 percent, which can be enhanced into a range of 18 to 23 percent by 2012, just by laser-based processes. Solar panel efficiency without sunlight concentrators will be in the range from 15 to 19 percent by the year 2012. Conversion efficiencies in the order of 23 to 25 percent are possible from high performance, high efficiency monocrystalline solar cells incorporating optimal design principles. Optimal design principles include highly effective antireflection coatings, narrow metal contacts, highly doped emitter regions, techniques to reduce recombination losses, excellent surface passivation, and back surface contacts with reflective properties.

4.5.2 Fabrication Steps Using Laser Technology

Maximum benefits of laser technology can be achieved in the manufacturing of solar panels on mass scale production. These benefits include higher number of

Table 4.3 Major Benefits Expected from the Implementation of Laser Technology for the Production of Crystalline Silicon Solar Cells

Measuring Parameters	Current Technology	5-Year after Laser Technology
Solar cell efficiency (%)	14–18	Increase by 3–4
Silicon wafer size (in.)	5 or 6	Close to 8
Wafer thickness (µm)	220	<200
"Green" production	Production-based	Cleaner production
Device yield level (%)	85–90	>95

wafers produced per hour, consistency in quality control, minimum production cost, high device yield, and uniformity in thin-film patterning. c-Si solar panel fabrication is divided into two stages to maintain quality control and device yield. Solar cells typically 125 or 150 mm wide and 220 mm thick are manufactured in a "back end" stage. The "front end" stage involves connecting 60 to 80 solar cells in series to form a solar panel with 200-W power rating. Laser processing steps are widely used during the "back end" stage fabrication. It is important to mention that the front-surface scribing on crystalline silicon can produce grooves that are roughly 30 µm wide and 35 µm deep, which are suitable for buried contact formation. Multiple grooves can be scribed on c-Si solar cells at about 100-µm pitch using a Q-switched diode-pumped solid-state laser and a scanning device. It appears that despite their advantages, laser processing is not being used overwhelmingly in solar panel production. Major benefits of laser processing technology can be accessed only after 5 years or so, when laser technology is fully implemented in the production of the solar panels. Major benefits or changes shown in Table 4.3 can be expected after 5 years of implementation of laser technology for the c-Si solar cell production.

Based on benefit projections, it can be stated that laser processing technology can help the solar industry to continue to grow at a faster rate, with significant improvement in conversion efficiency, device yield, and unit production cost. The principal benefits of the laser processing technology are described below.

4.5.2.1 Lasers Offer "Green" Technology

Greener processing is the greatest advantage, which is only possible from the deployment of laser technology, particularly when compared with screen-printing/lithographic technologies currently used in the manufacture of solar cells and panels. Since solar energy offers a low carbon footprint solution to future energy requirements, it is not surprising that priority is given to processing techniques with minimum environmental impact. Note a diode-pumped solid-state laser has an inherent advantage over other technologies such as etching solutions, which use toxic chemicals and produce hazardous waste and greenhouse effects. Furthermore, the idea of "green" laser-based processing procedures immediately resonate with current solar company mission requirements and marketing campaigns for boosting of solar technology.

4.5.2.2 Laser-Based Technology Best Suited for Thinner Wafers

Production of thin wafers with minimum cost is only possible with laser-based technology. It is important to mention that the current dominant cost within the c-Si solar cell manufacturing is the silicon raw material used at the "back end" stage. Therefore, reducing the silicon wafer thickness of 220 μm currently used by the solar industry can be significant cost savings. In addition, processing cost can be further reduced by moving to larger-area wafers. However, thinner and larger wafers are highly fragile and mechanically vulnerable, leading to lower product yields regardless of any form of contact processing. Furthermore, increased yields are highly essential for the solar industry involved in the manufacture of solar cells and panels at minimum cost. Note the noncontact nature of the laser processing provides the major benefit versus any contact-based alternative.

4.5.2.3 Edge Isolation Is the Most Critical Part of c-Si Production Lines

The solar cell industry is being driven increasingly toward the use of lasers and associated equipment to develop tools specially optimized to produce solar cells and panels for mass scale manufacturing. The laser technology is best suited for production of c-Si devices to optimize a technique known as "edge isolation." Edge isolation is an essential issue for all c-Si cell production lines; the front and rear surfaces of the silicon cell must be electrically isolated after a phosphorous diffusion process that dopes the p-type silicon with an n-type top layer, as illustrated in Figure 4.2.

Note only laser can scribe isolation grooves, which are typically 10 to 20 nm deep, to initiate shunt pathways between the front and rear surfaces. These narrow grooves are essential to achieve high electrical isolation. Furthermore, the grooves must be scribed as close as possible to the edges of the front surface to achieve maximum performance, which is only possible with laser operating at optimum wavelength. Although other etching technologies such as plasma and wet chemical technologies can compete with laser technology, nevertheless, laser technology offers optimum processing speed, groove uniformity, precision penetration depth, and excellent quality control. These benefits require optimum laser source and appropriate laser-based tooling to achieve higher edge electrical isolation.

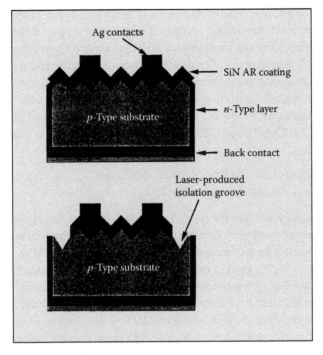

Definition of symbols:
   SiN stands for silicon nitride
   AR stands for anti-reflection
   Ag stands for silver

Figure 4.2   Architecture of a crystalline silicon solar cell showing the contacts, antireflection coating, and isolation grooves.

4.5.2.4 Laser Types and Performance Parametric Requirements

Studies performed by the author on diode-pumped solid-state (DPSS) lasers [5] are best suited for precision cutting and welding operations. These DPSS lasers offer higher reliability, excellent beam quality, high frequency stability, and optimum accuracy in cutting, marking, and welding operations. These lasers yield excellent brightness and high average and pulsed power outputs with impressive capabilities for soldering the leads in delicate electronic components and surface-mount devices, brazing operations and successful marking, and precision cutting/drilling operations. The studies further indicate that shorter penetration depths are only possible with lasers operating at ultra-violet (UV) wavelengths, which allow narrower grooves to be scribed that are needed to minimize the "dead" area around the edges and to maximize the conversion efficiency of the solar cell. Significantly higher absorption occurs in c-Si semiconductor medium at shorter wavelengths such as 355 or 532 nm. Silicon absorption is about four to five orders of magnitude stronger at 355 nm wavelength compared with a wavelength of 1064 nm, thereby allowing highly localized front-surface scribing, when a Q-switched UV DPSS is used. The UV wavelengths offer shorter penetration depths and narrower grooves in the c-Si semiconductor material, leading to high device yield and conversion efficiency of the solar cell with minimum production cost for the solar panel. In brief, grooves as narrow as 20 μm and as deep as 15 μm can be scribed in c-Si solar cells using a 355-nm Q-switched DPSS laser with average power less than 25 W.

4.5.2.5 Impact of "Microcracks" on Solar Cell Reliability and Yield

There is a high probability for the formation of "microcracks" at the cell surface at 1064-nm laser processing. According to research scientists, formation of microcracks is the key limiting factor, because they decrease the device yield of the production lines leading to higher production costs. Furthermore, any microcrack can result in solar cell failure during the remaining back-end and front-end production stages during the panel installation phase or under stress from weather-related conditions such as heavy rain or heavy snow or strong winds. It is absolutely essential to reduce the frequency of microcracks in the c-Si devices to preserve both high device yield and conversion efficiency. Industry experts believe that microcracks can be minimized by reducing the heat-affected zone produced by the 355-nm laser source. Reduction in the heat-affected zone is possible by keeping the average power output of the Q-switched DPSS laser below 25 W. Power scaling to tens of watts of average power with beam quality less than 1.3 combined with UV processing speeds more than 600 nm/sec represents the state of the art in laser-based edge isolation technology.

Laser-based tools offer technical advantages over competing technologies, processing more than 3000 wafers per hour. This particular wafer production rate is fully aligned to meet the needs of the solar cell industry at least over the next 5 to 10 years. Research and development activities are currently pursued by laser suppliers to develop alternate DPSS lasers operating at infrared, green, and other UV wavelengths with similar output power levels, but with nano- and picosecond pulse-width operation. The new generation of DPSS lasers will ensure much higher conversion efficiency, enhanced device yield, and higher reliability vital for faster solar industry growth.

4.6 Three-Dimensional Nanotechnology-Based Solar Cells

The principal objective of a three-dimensional (3-D) solar cell is to trap the incoming photons until they are absorbed and effectively converted into electricity. This design concept differs from the traditional planar solar cell approach, where photons have only one chance to interact with the conventional one-dimensional (1-D) solar cells. In the case of a 1-D solar cell, some photons are absorbed, but most of them are reflected by the flat surface and thus wasted, leading to very low theoretical conversion efficiencies, less than 15 percent for silicon cells.

4.6.1 3-D Solar Cells Using an Array of Carbon Nanotubes (CNTs)

Scientists are currently working on solar cells incorporating MEMS technology and nanotechnology concepts. A solar system using a carbon nanotube (CNT) array [6] coated with the semiconductor materials cadmium telluride (CdTe) and cadmium sulfide (CdS) is under development. The object is for the photons to become ensnared in the tiny chasms between the nanotubes until they are fully absorbed [6]. Laboratory tests performed on planar solar cells reveal a typical current density of 0.7 mA/cm². The 3-D solar cell using a CNT approach has demonstrated a current density of 44.4 mA/cm², which represents a 63-fold improvement.

This nanotechnology-based 3-D solar cell is fabricated by a molecular beam epitaxy method. The device measures 100 by 40 by 40 µm and the CNTs are separated by 10 µm as illustrated in Figure 4.3. This design concept when applied to solar cell manufacturing offers smaller arrays, compact size, and minimum production cost. Additional research and development efforts are required to determine the optimum dimension for the CNT elements and it will take at least 5 years of sustained research activity before this particular technology will be available for manufacturing of 3-D solar cells. Research scientists predict theoretical conversion efficiencies in the 20 to 25 percent range. However, cost and reliability data on these will not be available even after 5 to 7 years.

4.6.2 Solar Cell Design Configurations Using Nanowires, Nanocrystals, and Quantum Dots

Scientists at the University of Minnesota have designed photovoltaic solar cells using nanowires, nanocrystals, and quantum dots. Semiconductor nanocrystals are also known as quantum dots. It is important to point out that quantum dots have distinct advantages over photosensitive dyes in the fabrication of solar cells. Furthermore, quantum dots have a superior ability to match the solar spectrum, because their absorption spectrum can be tuned with the particle or dot size. Quantum dots have demonstrated a capability to generate multiple electron-hole pairs per individual photon, which could result in enhanced conversion efficiency for a solar cell. Research scientists believe that deployment of quantum dot technology in fabrication of solar cells will meet both critical requirements, namely, higher conversion efficiency and low fabrication cost.

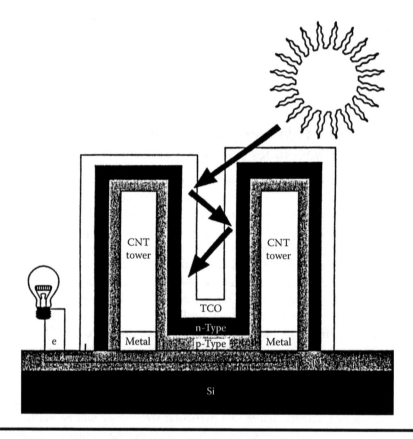

Figure 4.3 Architecture of a high-efficiency solar cell using an array of carbon nanotubes.

A solar cell can be fabricated with cadmium selenide (CdSe) quantum dots, which are attached to the surface of single-crystal zinc oxide nanowires. When this device is illuminated by visible light at 55-nm laser, quantum dots typically 2 nm in diameter inject electrons across their interface with the nanowires. These nanowires with lengths ranging from 2 to 12 μm provide a direct electrical pathway to the cell's photoanode, which has improved electron transport properties. Note these nanodots are grown directly into the substrate. Short-circuit current densities ranging from 1 to 2 mA/cm² and open-circuit voltages ranging from 0.5 to 0.6 V are generated in the nanowire/nanocrystal device when illuminated with a solar radiation intensity of 100 mW/cm². Internal quantum efficiency as high as 65 percent has been observed by research scientists, which is close to what similar nanowire dye-sensitized solar cells offer. However, in terms of conversion efficiency, the quantum dots-based device demonstrated an efficiency less than 3 percent compared with an efficiency of 12 percent for dye-sensitized solar cells. Scientists are working on alternate architectures and are considering suitable materials for quantum dots and nanowires to achieve higher conversion efficiencies. In addition, integration of nanotechnology-based zinc oxide (ZnO) nanorods will significantly improve the conversion efficiency of this particular device, but rough efficiency estimates are not available. According to the solar cell scientists, it will take at least 7 to 10 years before quantum dots-based solar cells become commercially available.

4.6.3 Multijunction Amorphous Nanotechnology-Based Solar Cells

The latest research studies indicate that an improved triple-junction amorphous solar cell comprising three distinct semiconductor layers with optimum thicknesses can achieve theoretical conversion efficiencies exceeding 35 percent by deploying solar concentrators [7]. This triple-junction amorphous nanotechnology-based solar cell, shown in Figure 4.4, consists of three semiconductor material layers, namely, the gallium indium phosphide (GaInP) layer, best suited for the short-wavelength region of the solar spectrum, the gallium indium arsenide (GaInAs) layer, best suited for the middle region of the spectrum, and the germanium (Ge) layer, which captures most energy from the infrared region of the spectrum. The subcells each defined by its semiconductor material layer are stacked one over another as demonstrated in Figure 4.4. Typical characteristics of photon-detector materials are summarized in Table 4.4. It is important to mention that significant improvement in conversion efficiency of this triple-junction solar cell comprising three different semiconductor layers is due to optimum absorption or extraction of the incident

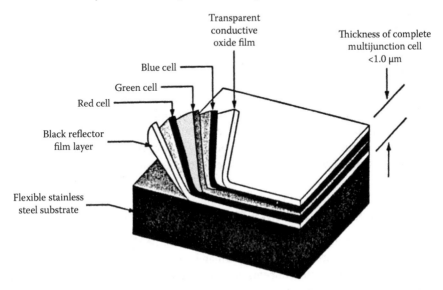

Figure 4.4 Triple-layer amorphous solar cell design using MEMS and nanotechnology concepts.

Table 4.4   Typical Characteristics of Photon-Detector Materials

Detector Material	Spectral Range (nm)	Responsivity (A/W)	Current Gain Ratio
Germanium (Ge)	40–1800	0.2 @ 400/0.8 @ 800 nm	200:1
		0.8 @ 1700/0.9 @ 1800 nm	
InGaAs	1000–1700	0.6 @ 1000/ 0.8 @ 1700 nm	10:1
GaAs	800–900	0.8 @ 850 nm	20:1

solar energy over the wide spectral region ranging from 400 nm to 1800 nm. The construction of this particular solar cell is based on a metamorphic principle known as lattice mismatch, which allows energy bandgaps to be manipulated so that each layer or subcell is more efficiently matched to a specific spectral range leading to higher absorption rate. These multijunction amorphous solar cells were developed specifically to provide electrical power for spacecraft or satellite applications. Research scientists believe that integration of solar concentrators will make these devices more affordable and will be best suited mostly for terrestrial applications. The scientists further believe that a solar power system using such cells can deliver electrical power in excess of 35 kW, almost 50 percent more than could be achieved with silicon-based concentrated solar cells.

Research scientists at Spectrolab Inc. in Sylmar, California, a subsidiary of Boeing Co., are experimenting with four-junction [7] amorphous solar cells using various metamorphic materials, other cell architectures and exotic design approaches, which could achieve conversion efficiencies close to 45 percent. Currently, four-junction cells with solar concentrators are showing conversion efficiencies exceeding 36 percent and the company claims that such solar cells with efficiencies close to 42.1 percent will be available by 2012.

These detector performance characteristics are specified at room temperature (300 K). Note multijunction solar cells are very complex and expensive, thus, their immediate deployment in solar power systems for homes and offices cannot be justified. As stated earlier these solar cells are specially developed for terrestrial spacecrafts and reconnaissance satellites, which require electrical power in excess of 30 kW.

4.7 Solar Concentrators for Efficiency Enhancement

As stated earlier, solar cell conversion efficiency can be significantly improved by using solar concentrators, which will ultimately reduce the solar electricity generation cost depending on the concentrator cost and concentration factor. Currently, solar-based electricity cost estimates vary from 23 to 32 cents per kilowatt-hour or per unit, while the residential utility-based electricity prices range from 5.8 to 16.7 cents per kWh. However, with current technology, "concentrated solar power"

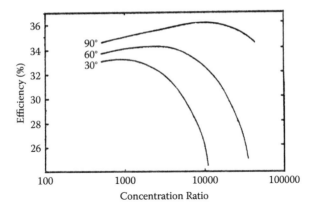

Figure 4.5 **Estimated conversion efficiencies of a silicon solar cell as a function of sunlight concentration ratio and acceptance half-angle for the incident sunlight.**

could cost about 40 percent less than the PV-based solar power system, which is close to the prevailing electricity prices in California.

Studies performed by the author reveal that there is some limitation on the conversion efficiencies of silicon solar cells when exposed to very highly concentrated solar light. This limitation is due to loss of conductivity modulation and loss in the carrier collection efficiency due to Auger effects under highly concentrated sunlight conditions. The studies further reveal that both the optimum base thickness of the solar cell and acceptance half-angle have an impact on efficiency upper bounds. The efficiency upper bounds as a function of solar concentration ratio or factor and acceptance half-angle are clearly visible in Figure 4.5.

The studies further indicate that the Auger recombination places the most severe intrinsic bounds on the conversion efficiency of silicon solar cells even under one-sun operating conditions. Furthermore, these bounds become more restrictive as the intensity of the incident sunlight increases. However, these bounds are found under maximum possible terrestrial sunlight concentration conditions. The restriction on efficiency bounds is strictly due to loss of bulk conductivity modulation and carrier diffusion lengths.

The impact of the cell's base thickness and acceptance half-angle (θ) on the conversion efficiency cannot be underestimated. Optimum base thickness requirements as a function of acceptance half-angle are summarized in Table 4.5.

4.7.1 *Impact of Base Thickness of the Solar Cell on Conversion Efficiency*

Solar light trapping in a photovoltaic cell or solar cell becomes more effective as the acceptance half-angle of the solar cell decreases. Furthermore, for each

Table 4.5   Optimum Base Thicknesses Compatible with Various Half-Angles

Acceptance Half-Angle θ (Degrees)	Optimum Base Thickness (μm)
15	15
30	30
60	60
90	100

acceptance half-angle, there is an optimal base thickness of the solar cell with limiting efficiency performance at any given sunlight concentration ratio. Note at low concentration ratios below 1000:1, the optimum thickness is virtually constant regardless of the concentration ratios. In other words, the optimum thickness is independent of sunlight concentration in low concentration environments. Estimated values of optimum thickness as a function of half-acceptance under high concentration factors are shown in Table 4.5. These values can be obtained using appropriate solar cell simulation programs. Solar cell parameters can be selected to model a lightly doped base region with Auger recombination limited lifetimes, which can be bounded by highly doped regions of opposite dopant types terminated by zero surface recombination velocity surfaces. The doped regions must be selected very thin so as not to contribute significantly in recombination in the solar cell structure.

Carrier-to-carrier scattering must be included in the modeling analysis. However, this will reduce the carrier mobility at high carrier concentration, which will slightly reduce the conversion efficiency below the calculated values obtained from computer simulation. Efficiency curves shown in Figure 4.5 indicate that the conversion efficiency begins to decrease significantly after the concentration ratio exceeds 1000:1 or 1000 suns. As mentioned earlier this efficiency reduction is strictly due to a loss in conductivity modulation near the rear contact of the solar cell.

4.7.2 Impact of Sunlight Concentration Ratio on Other Performance Parameters of the Solar Cell

Other solar cell performance parameters such as open-circuit voltage (V_{oc}), short-circuit current density (J_{sc}) and the fill factor (FF) are influenced as a function of sunlight concentration ratio (SCR), which in turn also impacts the conversion efficiency. Maximum change in open-circuit voltage for a 100-μm thick solar cell with a light acceptance half-angle of 90° occurs when SCR approaches 1000 and, thereafter, the increase in voltage is hardly 25 mV regardless of its value from 1000 to 40,000. It is important to mention that there is no change in voltage once the SCR attains a value of 20,000 and beyond. The short-circuit current in the same device continues to increase with increase in SCR. Note the sublinear increase in short-circuit current is a resistive effect due to the loss of bulk conductivity modulation. As far as fill factor is concerned, it is about 0.88 at SCR equal to one or at 1 sun, 0.81 at 1000 suns, 0.79 at 20,000 suns, and remains constant thereafter.

4.7.3 Optimum Cell Thickness

Since the above performance parameters of a solar cell are dependent on the cell thickness, it is desirable to determine the optimum thickness of the cell. It is important to mention that the optimum cell thickness is dependent on sunlight concentration ratio (SCR) and cell acceptance half-angle. It is stated in several technical papers on solar cells that thinning the solar cell leads to higher conversion efficiency. The analytical data summarized in Table 4.6 highlight the requirements for very thin silicon cells for high-efficiency performance under very high SCRs. Note due to loss in conductivity modulation at higher concentration ratios, the optimum cell thickness decreases with the increase in concentration ratio to values below 20 μm. However, overestimation of conversion efficiency bound at higher concentration ratios and the cell thickness has been observed in some published reports, which could be misleading.

The data summarized in Table 4.6 reveals that the thickness of a silicon solar cell must be reduced under smaller acceptance half-angles and when exposed to sunlight concentrated ratios exceeding 1000, if higher conversion efficiency is the principal design requirement. However, extremely thin wafers may not provide higher mechanical integrity and reliability under harsh thermal and mechanical operating environments, which usually prevail in space environments.

Table 4.6 Optimum Thickness (in μm) of a Silicon Solar Cell as a Function of Sunlight Concentration Ratio and Cell Acceptance Half-Angles (θ)

SCR (Suns)	$\theta =$		
	90°	60°	30°
1	100	60	30
10	100	60	30
100	95	60	30
1000	65	54	30
10,000	39	24	11

Figure 4.6 **A highly efficient solar power system comprising bifacial solar modules and hemispherical compact mirrors, resulting in more electrical power output per module, with large space saving.**

4.8 Solar Cells with Specific Shapes and Unique Junction Configurations to Achieve Higher Performance

In this section, solar cells with specific shapes and junction configurations capable of providing enhanced performance capabilities will be discussed briefly. Critical performance parameters are identified, with major emphasis on spectral efficiency, conversion efficiency of the cell and the power output of the solar module.

4.8.1 Benefits of Bifacial Solar Modules

Research studies indicate that solar modules employing bifacial solar cells are very light sensitive on both sides of the PV devices. A photovoltaic solar power system comprising bifacial solar modules and hemispherical mirrors is shown in Figure 4.6. This arrangement of bifacial solar modules and hemispherical mirrors offers both the higher module conversion efficiency and power output with minimum space occupied in the solar panel. Based on nonimaging optics, it is evident that all the solar radiation falling upon the aperture will be directed onto the vertical solar modules enclosed by the semicircular mirrors. The principal advantage of this concept is that the area available on the roof for installation can be much better utilized. The second is that each solar module receives more radiation than a one-sided solar module would, leading to higher power output per module in an optimized arrangement on the solar panel. Under identical conditions, it is estimated that the bifacial module offers 1.6 times more electrical power output, even after including the reflection losses. It can be mentioned that under first approximation the solar collectors are independent of orientation and, thus, can be adapted to the shape of the roof, which offers significant advantage in the installation of solar panels on the roof. However, there is one slight disadvantage with this concept, that irradiation at the solar modules is not uniform in the vertical direction. Since there is uniformity parallel to the horizontal plane, this effect can be minimized by connecting the horizontally based solar cells in series. This concept yields maximum economy because the expensive module area is replaced by relatively cheap mirrors and offers lowest area of the solar power source.

4.8.2 Performance Enhancement from a V-Shaped Solar Cell

Most of the photons that hit the surface of low-cost, thin solar cells simply bounce off the surface, not allowing their energy to be captured, leading to poor conversion efficiencies of the solar cells. The photon energy can be captured by changing the basic architecture of the solar cell so that the photons can be trapped when they come in contact with the cell surface, which will provide greater chance for absorption needed to convert into usable electrical power.

One of the most efficient light-trapping techniques involves the use of metal gratings consisting of buried nanoelectrodes and scattering elements to enhance sunlight absorption. But these techniques are expensive and their performance degrades over time. A most efficient as well as inexpensive sunlight trapping design approach is illustrated in Figure 4.7. The geometry of V-shaped solar cell design allows optimum sunlight trapping by acting as an optical funnel. As a result, the solar photons have multiple interactions with the solar cell structure. This solar cell architecture permits a very thin active layer of the cell compared with the substrate, which is highly essential for high quantum and conversion efficiencies. The thin-film cells are outfitted with a zigzag pattern, with saw-tooth structural design. Note the repeating V-shaped design is considered to be very simple and cost effective and offers maximum quantum efficiency for all incident angles. A similar approach can be used with thick-film solar cells, but at the expense of lower conversion efficiency. The active layer of the cell and the reflective metal electrode are deposited onto a V-shaped transparent substrate electrode as illustrated in Figure 4.7. The V-shaped fold acts like an optical funnel, which draws the photons into the tiny cavities and causes them to bounce multiple times against the walls of the solar cell. Note the density of the reflections increases as the opening angle decreases. Each photon has

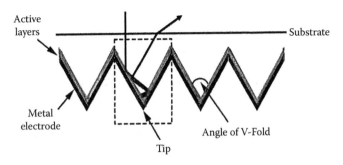

Figure 4.7 Geometrical configuration of a V-shaped cell capable of trapping solar light by serving as an optical funnel, thereby forcing the photons into multiple interactions within the solar cell.

multiple interactions with the cell, instead of a single interaction, which has been observed with a traditional planar solar cell.

Because of multiple interactions with the solar cell, the V-shaped solar cell has higher external quantum efficiency over the spectral region ranging from 400 to 800 nm and remains effective regardless of angle of incidence. The maximum quantum efficiency requires a V-fold design with a 35° angle. Preliminary calculations indicate that a 170-nm thick V-shaped solar cell can provide a 52 percent improvement in the quantum efficiency, when the V-fold design uses a 35° angle. There is one disadvantage of this cell design, that it requires more active material for its fabrication than the traditional planar solar cell. However, this can be compensated for by the improved conversion efficiency and power output capability of the V-shaped device.

4.8.3 Tandem Junction Cell

The tandem junction cell (TJC) is essentially a high-performance silicon solar cell, which is best suited for terrestrial solar power systems. The most distinctive design feature of the TJC device is the use of only back contacts to eliminate the metal shadowing effects. It provides efficient and cost-effective interconnection capability. Operation and performance of this cell can be explained by modeling a transistor action. The design relationships for conventional solar cells do not apply to TJC structures. However, excellent performance can be obtained by empirical optimization. A conceptual model can provide reliable information on the device operation and general design considerations and should provide a foundation for a rigorous computer analysis to predict critical performance parameters.

4.8.3.1 Modeling of TJC Parameters

Modeling of critical performance parameters of a TJC device can be accomplished using the structural details of the emitter, base, and collector elements of the device illustrated in Figure 4.8. Conventional TJC cell structure using flat-plate configuration can yield conversion efficiency close to 16.5 percent under AM1 spectrum [8]. However, a concentrator TJC solar cell has demonstrated an efficiency very close to 17 percent at sunlight concentration factor of 20, which is equal to 20 sun with AM1. The characteristic structure of the TJC device illustrated in Figure 4.8 clearly shows the front illuminated side of the texturized silicon surface with a thin active junction consisting of interposed n^+ and p^{+n} fingers at the back surface of the device. Note the refraction of the light within the silicon medium increases the optical path length and causes the light to strike the back surface at greater than the critical angle of 15° for total reflection, leading to a high percentage of light absorption in very thin TJC cells. The TJC design with back contacts offers high collection efficiency because of the thin base region and base material with long lifetime.

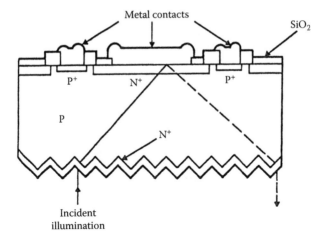

Figure 4.8 Structural design aspects of a tandem junction solar cell showing critical elements of the cell, including metallic contacts and various regions exposed to incident illumination.

The cross-section of the TJC device is similar to that of a transistor model, as shown in Figure 4.8. The TJC cross-section and equivalent transistor circuit are clearly shown in the figure. It is important to mention that the front n^+ corresponds to the emitter of the transistor, the p region is recognized as the base of the transistor and the back n^+ is designated as the collector of the transistor. As in a conventional transistor, the equivalent circuit identifies all the critical parameters of the TJC device. The current source is due to the hole-electron pairs generated in the emitter or in the base section near the emitter. The model can be used to describe the current collection for the short-circuit condition and then the open-circuit voltage. As shown in Figure 4.8, the minority carriers or holes generated in the front n^+ region (emitter) diffuse to the emitter-base junction and ultimately are swept by the electric fields into the base regions. Specific details on carrier generation in emitter and base regions under the incident illumination conditions are shown in Figure 4.9. A forward-bias voltage is generated across the junction such that electrons are injected into the base region in roughly equal quantities, which will yield high injection efficiency. It is important to mention that a voltage is needed for the injection of electrons back into the base regions. The output of the device is between the collector and base terminals.

The equivalent circuit parameters of the transistor are characterized by the well-known Ebers–Moll model [8]. The critical parameters characterized by this model are:

■ Current transfer ratio under forward (normal) bias conditions.
■ Current transfer ratio under inverse bias conditions.
■ Saturation current for the collector-base junction.

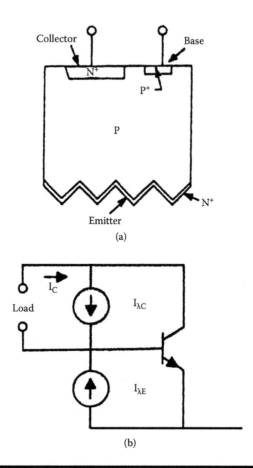

Figure 4.9 TJC device configuration showing transistor model of the cell: (a) Cross-section; (b) equivalent circuit; and carrier generation in the (c) emitter and (d) base regions. *Continued*

An important assumption of the Ebers–Moll model is that high electron injection does not change emitter efficiency by the conductivity modulation of the base. It is evident from the TJC device parameters shown in Table 4.7 that the effective base concentration is changed less than 4 percent for insolation at AMO condition. The relationships defined by this model are strictly valid for ideal diodes; that is, junctions with no recombination or leakage current. However, this model has been extremely useful for electrical performance analysis of practical transistors or TJC solar cell devices. Furthermore, this model will be found extremely valuable as a design guideline for the TJC solar cell devices.

Both the short-circuit current (I_{sc}) and the open-circuit voltage (V_{oc}) can be computed using this model. The dark current of the TJC is an important parameter

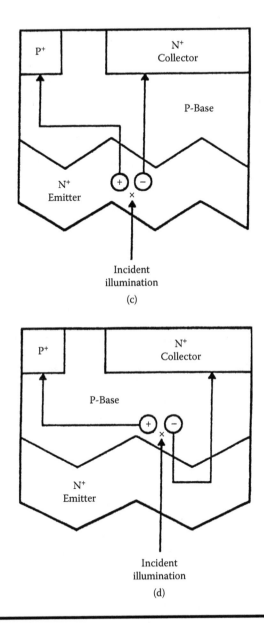

(c)

(d)

Figure 4.9   *Continued.*

Table 4.7   Critical Design Parameters for Various Regions of a TJC Solar Cell

Specific Device Region	Typical Dimensional Parameters
Base region (p-type)	Resistivity: 6 ohm-cm
	Base thickness: 110 µm
Emitter region n⁺ (front surface)	Junction depth: 150 nm
Collector back surface region (n⁺)	Junction depth: 600 nm

and is also known as the collector-base saturation current (*Icbs*). The expression for the open-circuit voltage of a TJC device can be written as

$$V_{oc} = [kT/q] \; [\ln(I_{sc}/I_{cbs})] \tag{4.6}$$

where k is the Boltzmann constant, T is the ambient temperature (K) and q is the electronic charge. High open-circuit voltages from TJC devices are possible at low recombination in the base region and at the surfaces. In addition, higher spectral response and collection efficiency are strictly possible under appropriate bias conditions.

4.8.3.2 Design Considerations for Optimum Cell Performance

Design parameters and operating biasing conditions capable of yielding optimum device performance are considered in this section. Well-established design procedures for DC characteristics of a semiconductor transistor are available in textbooks on transistors published several decades ago. One can acquire an expression for current transfer, which is a function of injection efficiency and the transport factor.

For an emitter junction, as shown in Figure 4.9, the injection efficiency can be defined as the ratio of injected electron current to total emitter current. Note the injection efficiency is dependent on the impurity profile and the processing technology used for the emitter region. Similarly, the injection efficiency for the inverse bias operation (i.e., when the collector is biased as an emitter) depends on the properties of the collector region. First-order theory indicates that the injection efficiency is increased by the use of heavily doped emitter regions. However, any effective shrinkage of the bandgap could lower the injection efficiency.

Recent studies reveal that the injection efficiency for the front junction of the TJC device should be high since the metal contacts are omitted. Deep junctions can be used at the back of the device for injection efficiency, because the generation rate at the back surface is relatively low.

The transport factor is strictly dependent on the inject electron current. However, the transport factor is the fraction of the injected electron current that reaches the collector-base junction. The reduction in the injected current in the base transit process is due to the bulk and recombination process. Note the bulk recombination is reduced at low ratios of bandwidth to diffusion length (*L*). This diffusion length is defined as

$$L = [\sqrt{D\tau}] \tag{4.7}$$

where *D* is the diffusion coefficient and τ is the minority lifetime in the base region of the device. Note bulk recombination can be minimized by using a high-resistivity base material for which high values of these parameters can be obtained. Surface recombination is strictly due to the p and p⁺ regions shown in Figure 4.8 and can be reduced by using smaller contact areas.

4.8.3.3 Projected Performance Parameters of TJC

The device performance is dependent on the collection of carriers from the contact-free front n⁺ region, as illustrated in Figure 4.8, and the geometrical dimensions of the device. Performance is also severely affected by the spectral response (mA/mW) and the effectiveness of the antireflection coating as a function of wavelength. Multilayer antireflection (MLAR) coating is necessary for minimum reflection loss at short wavelengths ranging from 0.4 to 0.6 μm. Note the spectral response is contingent on the light bias levels, which improves with the increase in light bias level. A steady-state light bias source is essential, if optimum performance is the principal requirement. The quantum efficiency of the TJC device is the most critical performance parameter, which is an indicator for the conversion efficiency of the device.

As far as the impact of dimensional parameters on the TJC device performance is concerned, quantum efficiency for the MLAR cell is better than 70 percent for illumination at 0.4 μm. As far as photon absorption capability is concerned, more than 65 percent of the total photons in the n⁺ region are absorbed at a wavelength of 0.4 μm in a TJC device having a junction depth of 150 nm. A higher percentage of photon absorption is possible with thinner junction depths, provided other performance parameters are not adversely affected. Note the collection efficiency cannot exceed 36 percent unless there are additional carrier collections from the n⁺ emitter region, as shown in Figure 4.9.

Both the current level and the current transfer ratio can affect the collection efficiency, which will impact the cell conversion efficiency. The light level from a steady-state laser source increases the current level and the current transfer ratio. The conventional TJC device has demonstrated an open-circuit voltage as high as 0.615 V under AM0 condition, while a back surface field TJC cell demonstrated the highest voltage of 0.622 V under AM1 condition. Note high open-circuit voltages are possible with low combination in the base region and at surfaces and with reduction in dark current levels. Quantum efficiency projections for a TJC device are shown in Table 4.8.

Table 4.8 Quantum Efficiency Projections for a TJC Solar Cell

Wavelength (μm)	Quantum Efficiency (%)
0.4	80
0.6	85
0.8	91
1.0	86

Table 4.9 Effects of Light Bias as a Function of Wavelength on the Spectral Response of a TJC Solar Cell

Wavelength (μm)	Spectral Response (mA/mW)	
	Without Bias	With Bias
0.4	0	0.25
0.6	0.12	0.43
0.8	0.37	0.62
1.0	0.51	0.70

These quantum efficiency projections indicate that highest efficiencies are possible over a spectral region ranging from 0.6 to 1.0 μm. Lower quantum efficiencies are due to less photon absorption at lower wavelengths. The effects of light bias levels on the spectral response of a TJC solar cell are visible from the spectral data shown in Table 4.9.

The data indicates that under zero bias conditions, the spectral response is extremely poor, which will significantly reduce the conversion efficiency of the TJC solar cell at lower wavelengths. The improvement in the spectral response due to biasing is evident. However, biasing will require a steady-state laser source, which will introduce higher cost and complexity.

4.9 Summary

This chapter focuses on solar cell designs capable of yielding high conversion efficiencies with minimum fabrication costs. The solar cells described are best suited for domestic and commercial solar-based power systems. High-efficiency solar cells best suited for space applications are described in a separate chapter. Potential techniques capable of achieving significant improvements in conversion efficiency, fundamental collection efficiency and internal collection efficiency are discussed. Contact types and requirements for silicon solar cells capable of providing higher conversion efficiency and reliability over extended periods are summarized. Front radial contacts fabricated from Ti/Pd/Ag metals and using 80-nm thick SiO_2 antireflection coating seem to boost the conversion efficiency 3 to 4 percent. In addition, deposition of silver/aluminum paste as back contact, deposition of silver paste as front contact, and deposition of aluminum paste on the back contact will further enhance the conversion efficiency of the solar cell. Internal light intensity enhancement using texturized optical sheets will certainly improve the conversion efficiency of the solar cells. Transmittivity data of radiation incident into a flat surface of optimum refractive index indicates that intensity enhancement offers significant improvement in both the conversion and collection efficiencies. Conversion efficiency enhancement in silicon solar cells is possible under highly concentrated sunlight conditions as a function of cell base thickness and acceptance half-angles. However, some loss in conductivity modulation occurs at higher concentration factors; the optimum cell thickness is restricted to below 20 μm. Various solar cell design configurations such as V-shaped cells, V-groove multijunction (VGMJ) cells and tandem junction cells (TJC) are described, with emphasis on reliability, conversion efficiency, fabrication costs, and specific application. Performance of solar cells fabricated with materials such as nanoparticles, nanodots, and carbon nanotubes (CNTs) are described, with particular emphasis on conversion efficiency and fabrication costs. Benefits from the use of dielectric microconcentrators in solar cells are identified.

The tandem junction cell (TJC) is a high-performance silicon solar cell for potential applications in terrestrial solar power systems. The TJC will not be attractive for domestic and commercial solar power system applications because of lower conversion efficiency and steady-state bias requirement. Research scientists believe that 3-junction and 4-junction amorphous solar cells using exotic design techniques are capable of yielding conversion efficiencies in excess of 30 percent by 2010. It is important to mention that the requirements of high conversion efficiencies and minimum fabrication costs are difficult but not impossible. However, tradeoff studies may be necessary to meet both the performance objectives of solar cells in the not too distant future.

References

1. Perez-Quintana, A. Martel et al. "Comparative study of metal semiconductor contact degradation by current pulses on silicon solar cells with two contact types," *Solid-State Electronics* 45 (2001): 2017–2021.
2. P.R. Campbell and M.A. Green. "On intensity enhancement in texturized optical sheets for solar cells," *IEEE Transactions on Electron Devices* ED-33, No. 11 (1986): 1834–1835.
3. M.A. Greenwood. "Photocurrent altered with nanoparticles," *Photonics Spectra,* March 2008, 106.

4. Finlay Colville and Corey Dunsky. "Lasers support emerging solar industry needs," *Photonics Spectra*, April 2008, 44–47.

5. A.R. Jha. *Infrared Technology: Applications to Electro-optics, Photonic Devices and Sensors.* New York: John Wiley and Sons, Inc., 2000.

6. A.R. Jha. MEMS and Nanotechnology-Based Sensors and Devices for Communications, Medical and Aerospace Applications. Boca Raton, FL: CRC Press, 2008.

7. M.A. Greenwood, News Editor. "Solar Technology: Seeking its day in the sun," *Photonics Spectra*, July 2007, 42–50.

8. W.T. Matzen, S.Y. Chiang et al. "A device model for the tandem junction solar cell," *IEEE Transactions on Electron Devices* ED-26, No. 9 (1979) 1365–1368.

Chapter 5

Solar Cells Deploying Exotic Materials and Advanced Design Configurations for Optimum Performance

5.1 Introduction

Second- and third-generation solar cells employing exotic materials are described in this chapter, along with advanced design concepts capable of providing moderate to high conversion efficiencies but with minimum material and fabrication costs. Delay in deployment of solar cells on a mass scale is due primarily to the solar cell costs, which include the material cost and the fabrication cost. New design concepts and material requirements to obtain maximum conversion efficiency for the solar cell with minimum manufacturing costs will be given prime consideration. The conversion efficiency can be derived in two distinct ways, namely, thermodynamics and the detailed balanced principle. The thermodynamic efficiency contribution is limited by the Carnot cycle relation. If the boundary conditions for terrestrial conversion are taken into account, a maximum theoretical efficiency of

85 percent is possible. The detailed balanced principle is strictly based on balancing the different particle fluxes in the solar cell structure and this method yields very similar results to those provided by the thermodynamic limit. It is important to point out that the practical efficiencies available from solar cells are far less than the theoretical efficiency limit. Based on the above statements, the reasons for lower efficiencies are obvious. First, the solar spectrum is very broad, ranging from the ultraviolet to near infrared region, whereas a semiconductor material such as silicon or gallium arsenide can only convert photons with the energy of the bandgap with optimum efficiency. Photons with lower energy are not absorbed and those with higher energy levels are reduced to gap energy by thermalization of the photo-generated carriers. But this situation can be controlled or rectified by using several thin layers of the semiconductor in tandem cell configurations. Second, the solar radiation arrives at the surface of the earth in a very dilute form. The weak direct sunlight can be concentrated by optical mirrors, dielectric microconcentrators, and low-cost polished reflectors, leading to higher conversion efficiency. In addition, the responsivity of the material as a function of wavelength plays a key role in yielding optimum performance. Advanced solar cell design configurations will be investigated in future research and development activities with particular emphasis on efficiency, reliability, and manufacturing costs. Note some materials might yield high conversion efficiency, but their fabrication and manufacturing costs are not acceptable. For new materials, such as organic materials or those derived from nanotechnology, current efficiencies vary from 5 to 8 percent, but they can be fabricated with minimum costs. However, efficiency of these materials could improve in the next 5 to 10 years. Tradeoff analysis for the solar cells will be performed in terms of efficiency, design complexity, and manufacturing costs.

5.2 Potential Materials for Solar Cell Applications

Potential solar cell materials will be investigated for maximum efficiency and without any concentration schemes. Note the bandgap energy of the semiconductor material largely determines the efficiency of the solar device. Furthermore the physics of the materials do not solely establish the criterion for the selection of the semiconductor material. Cost and availability of the chemical elements come into play when the issue of mass production of solar cells is addressed. Considering all the pertinent issues, silicon is by far the most frequently used material for the production of solar cells. Gallium arsenide and indium are rarely used due to their higher cost and toxicity. Gallium arsenide solar cells are deployed in space applications, where cost is not a significant factor. Arsenic, selenium, and cadmium are dangerous. All of these considerations point toward a clear preference for the use of silicon in fabrication of solar cell devices.

Currently, material with complex chemical composition, such as cadmium indium deselenide (CIS), is receiving the most attention because of lower fabrications cost and less complex processing. Such solar cells with large areas can be manufactured with minimum cost using a straightforward co-evaporation technique. These solar cells do contain small amounts of toxic materials, but they are needed in small quantities because of large areas. Such solar cells have demonstrated conversion efficiencies exceeding 10 percent. Currently, organic materials for solar cell applications are being widely investigated, but their conversion efficiencies are approaching at best about 6 percent. Research scientists deeply involved in the design and development of organic solar cells predict that it will be difficult, if not impossible, to extend the conversion efficiencies beyond 10 percent. Solar cell designers and research scientists are considering electrochemical solar cell designs incorporating specific dyes.

Cost reduction demands simple design concept and least expensive fabrication processing. Japanese scientists involved in the design and development of amorphous hydrogenated silicon (a-Si:H) solar cells reveal that these cells offer low conversion efficiencies and are best suited for hand-held electronic devices requiring minimum electrical power consumption.

Scientists working on solar cells using microcrystalline silicon and polycrystalline silicon materials predict that these materials will provide cost-effective performance and with acceptable compromise in terms of efficiency and fabrication cost.

5.2.1 Critical Performance Parameters and Major Benefits of Materials

Performance parameters and major benefits of materials such as silicon crystalline materials, gallium arsenide materials, inexpensive dye-sensitized semiconductors, nanotechnology-based materials, organic materials, antireflection coating materials, PV-based photonic crystals, ternary compound semiconductor materials, including the copper indium diselenide (CuInSe)-based materials (i.e., CIS, CISG, CdS, and Cd Te), amorphous silicon (a-Si) materials, microcrystalline silicon (μc-Si) materials, nanocrystalline silicon materials (nc-Si), and amorphous hydrogenated silicon (a-Si:H) and hydrogenated nanocrystalline silicon (nc-Si:H) materials will discussed. Extensive research and development activities are currently being carried out on the design of solar cells using crystal silicon, thin-film silicon, amorphous silicon, organic, and CIS/CIGS materials. Table 5.1 summarizes the projected conversion efficiencies of solar cells using the above-mentioned materials [1].

5.2.2 Critical Properties Requirements of Semiconductor Materials

It is important to point out that the conversion efficiency of a solar cell is dependent on the angle of incident the sunlight makes with the device surface and the air mass

**Table 5.1 Future Projected Conversion Efficiencies of
Various Solar Cells as a Function of Timeframe (Percent)**

Material Technology	Year				
	1980	1990	2000	2010	2020
Crystal silicon	16	18	22	25	27
Amorphous silicon	8	12	15	17	19
Thin-film silicon	—	1	13	21	24
CIGS	5	13	16	22	23
Organic	—	—	3	8	12

(AM) numbers. AM refers to the intensity and the spectral distribution resulting from a certain path length of the sunlight through the atmosphere. AM1 means the sun is directly overhead (i.e., when the sunlight is normal to the cell surface). AM1.5 means that the sun is at such a position that the path length is 1.5 times more compared to that at AM1. In addition, the light absorption also plays a key role in the conversion efficiency of the cell, which is strictly dependent on the thickness of the cell base material. Light absorption is considerably weaker in an indirect semiconductor material than in a direct semiconductor material. For example, it takes only 1 μm of GaAs (a direct semiconductor material) compared to 100 μm of silicon (an indirect semiconductor material). From solid-state physics silicon is not an ideal material for photovoltaic conversion applications. Note silicon is considered as an indirect semiconductor material based on valance band maximum and conduction band minimum values.

Silicon is widely used in the design and development of solar cells, because silicon technology is highly developed and fully matured. It is produced in large quantities at relatively low costs compared to other materials used in fabrication of solar cells. Regardless of the disadvantages of this material, the future of solar energy materials must address the following issues to achieve low-cost material technology:

■ Continued dominance of the single crystal or polycrystal technology.
■ Development of new crystalline thin-film silicon materials with low and medium thicknesses in the form of either ribbons or thin substrates.
■ Acceleration to mass production for the latest true thin-film materials such as amorphous silicon (a-Si) or CIS or CdTe.

Major requirements for the ideal solar cell materials are summarized in Table 5.2.

Conversion efficiency of a solar cell is strictly dependent on the semiconductor bandgap at a given AM. Many semiconductor materials are available, which will

Table 5.2   Major Requirements for Ideal Solar Cell Materials

Material Requirements	Si	a-Si	CIGS/CdTe
Bandgap range from 1.1–1.7 eV	Yes	Yes	Yes
Nontoxic	Yes	Yes	No
Readily available	Yes	Yes	No
Direct band structure	No	Yes	Yes
Large area reproducibility and yield	No	Yes	Yes
Long-term stability	Yes	Yes	Yes
Good photovoltaic efficiency	Yes	No	Yes

increase the likelihood of achieving the goal of low-cost solar cells. Furthermore, the new classes of photovoltaic semiconductor materials such as organic solar cells or thin-film tandem cells will bring improved performance capabilities in the years ahead [1].

5.2.2.1 Amorphous Silicon (a-Si) Material

Technical papers and reports relevant to amorphous silicon solar cells first appeared in the late 1960s. Improvement in the performance of an a-Si solar cell has been very slow even after 25 years of research on this material. The conversion efficiency drops as the cell is exposed to light. The efficiency degradation is primarily due to reduction in fill factor and short-circuit current density, but the open-circuit voltage remains practically unaffected. The reason for the reduction in cell performance has not been fully explained. However, research scientists predict that the recombination of light-generated light carriers causes weak silicon-hydrogen bonds to be broken in the amorphous material, thereby introducing additional defects in the material that lower the collection efficiency and increase the series resistance. Seasonal variations on the conversion efficiency have been observed by German research scientists. The a-Si solar cell efficiency tends to drop in winter, but recovers in summer due to annealing effects. This means that the a-Si solar cell efficiency remains fairly constant in countries located near the equator. The bandgap of a-Si is better matched to the solar spectrum than that of the crystalline silicon material. This means that on a per watt basis amorphous silicon solar cells have higher electrical power output integrated over a one year period. Currently, applications of a-Si solar cells are limited to consumer electronics, but these cells will found applications in solar electrical power generation, once the problem of light-induced degradation is solved. Stabilized a-Si solar cell efficiencies are close to 13 percent under laboratory conditions, which may slightly decrease under field environments. The module efficiencies will be under 8 percent [1].

5.2.3 Efficiency Limitations Due to Properties of Material and Deposition Techniques

Sometimes there are efficiency limitations posed by the material properties and the film deposition techniques employed. For example, in the case of crystalline silicon material, the distribution of band length and bond angles disturbs the long range of the crystalline silicon lattice order, which changes the optical and electronic properties of the materials. In the case of amorphous silicon, the optical gap increases from 1.12 eV to 1.76 eV, which affects the efficiency of the solar cell as a function of AM and solar spectral wavelength. It is important to mention that amorphous silicon is an alloy of silicon and hydrogen and that is why it is sometimes referred to as amorphous hydrogenated silicon (a-Si:H). The semiconductor bandgap of pure a-Si:H material is about 1.6 eV and the maximum theoretical efficiency for thin a-Si:H solar cells varies approximately from 25 percent under AM1 to 27 percent under AM1.5 to 29 percent under AM0 conditions or black-body limitation. However, a maximum conversion efficiency of 15 percent has been observed for a-Si:H solar cells under laboratory conditions. Note the efficiency of commercial devices is between 8 and 10 percent, which can be improved using exotic vapor deposition techniques and thin-film materials with minimum impurities.

Techniques are available for deposition of thin films using silicon material for amorphous hydrogenated silicon (a-Si:H) or nanocrystalline a-Si:H (nc-Si:H) solar cells. These cells are best suited for diverse applications such as active matrix displays, image sensors, and solar cells with low electrical outputs. Potential deposition techniques [2] include conventional plasma enhanced chemical vapor deposition (PECVD), VHF-PECVD using VHF frequencies, hot wire chemical vapor deposition (HWCVD), pulsed-PECVD, and modified pulsed-PECVD. One must select the deposition technique that will offer minimum light-induced degradation effect, optimum deposition rate, improved cell efficiency, and high device yield. The deposition rate of 1 A/sec yields high quality a-Si:H films, but at higher production costs. Furthermore, film thickness of 1 to 3 μm is best suited for nanocrystalline nc-Si:H solar cells. Note both the deposition rate and the cell thickness determine the efficiency for the a-Si:H and nc-Si:H solar cells. However, a deposition rate of 1 A/sec is necessary to achieve high cell efficiency, when the device thickness varies between 1 and 3 μm.

5.2.4 Impact of Deposition Process on Cell Efficiency and Yield

Solar cells using a-Si:H thin films with 1 A/sec deposition rates have demonstrated conversion efficiencies less than 8 percent, leading to poor device yields and high production costs. Microcrystalline solar cells using PECVD, VHF-PECVD, and WHCVD deposition techniques offer conversion efficiencies in the range of 7 to 9 percent. However, the "micro-morph" solar cells comprising a-Si:H and nc-Sc:H materials using PECVD deposition techniques have demonstrated efficiencies close to 13 percent at a deposition rate of 1 A/sec. Solar cell devices using the VHF-PECVD technique resulted in a conversion efficiency close to 7 percent in a single-junction structure with a deposition rate of 5 A/sec. Efficiencies for PIN solar cell devices have been reported better than 9 percent using the HWCVD deposition technique, but with a deposition rate of 1 A/sec. Conventional PECVD techniques involving high pressure and low substrate temperature have shown conversion efficiencies close to 10 percent, but these process conditions are not suitable for solar cell production due to lower yields. Conversion efficiency for a single-junction nc-Si:H-based PIN solar cell (shown in Figure 5.1) is hardly 7.5 percent. All these deposition techniques so far discussed suffer from low deposition rates, dust formation, and the compatibility issue of large area deposition.

In summary, the pulsed-PECVD technique, which involves modulating the plasma at frequencies in the range of 1 to 100 kHz, offers suppression of dust formation, thereby increasing both the yield and conversion efficiency. State-of-the-art a-Si:H materials and solar cells can be produced successfully using the pulsed-PECVD deposition technology [2]. In addition, state-of-the-art nanocrystal silicon hydrogenated (nc-Si:H) films and single-junction solar cells based on nc-Si:H technology can be produced with minimum costs and complexity. Quantum efficiency improvement can be achieved using a modified pulsed-PECVD processing technology, which controls the optoelectronic properties of nc-Si:H materials and performance parameters of PIN solar cell devices.

5.2.5 Optoelectronic Properties of Nanocrystalline Silicon Materials

Optoelectronic properties of nanocrystalline silicon materials play a key role in the performance optimization of PIN solar cell devices and single-junction a-Si:H solar cells. Several factors determine the optoelectronic properties of nc-Si:H materials, such as the orientation and passivation of grains, oxygen concentration, processing temperature, and crystalline fractions. From a solar cell device point of view, the critical factors include minimization of the incubation layer, control of interfaces, minority carrier diffusion length, the thickness of nc-Si:H absorber intrinsic layer and the effects of textured substrate. A diffusion length of 1.2 μm is considered appropriate for optimum PIN solar cell devices.

A modified pulsed-PECVD deposition technique is most ideal to deposit a nc-Si:H absorber intrinsic layer on a "superstrate" type configuration, while doped layers can be grown using the normal continuous wave (CW) technique at a fixed frequency of 13.56 MHz for optimum device performance. Note PIN solar cell devices can be grown simultaneously on crystalline silicon wafer at different temperatures to observe spectral parameters. In case of nc-Si:H devices, nc-Si:H film

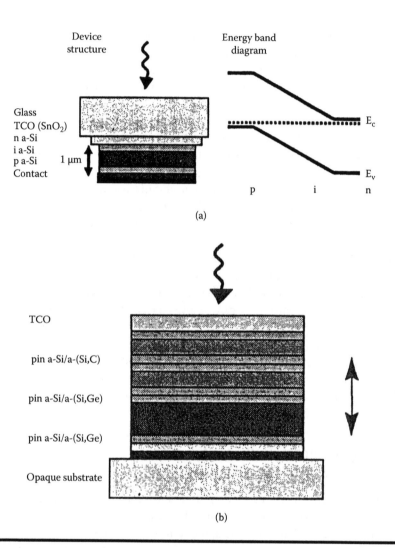

Figure 5.1 Design details for (a) an amorphous PIN solar cell and (b) a tandem solar cell using amorphous silicon technology.

consists of mostly oriented grains with crystalline fraction of 80 percent and with size ranging from 15 to 18 nm. A passivation reaction is necessary to achieve good passivation of the grain boundaries.

5.2.6 Impact of Various Interface Layers on the Performance Parameters of nc-Si:H-Based PIN Solar Cell

For superstrate-type PIN solar cells, the p-i interface plays a key role in determining the device electrical performance, as nc-Si:H often starts with an amorphous incubation phase. The sensitivity of the cell varies with the film thickness and growth conditions in terms of temperature and pressure. Note the electrical lateral conductivity of 0.3/ohm.cm is considered optimum for the nc-p layer, which acts as a dopant layer. This lateral conductivity increases several orders of magnitude in the thickness ranging from 20 to 30 nm. This radical increase in conductivity indicates the presence of an amorphous incubation phase in the nc-p layer deposited on the glass, as illustrated in Figure 5.1 (a).

The thickness of the i-layer determines the fill factor, which in turn determines the conversion efficiency. A fill factor (FF) of 0.68 or higher is considered optimum for the a-Si:H i-layer. This indicates that the doped layers (nc-p and a-n) are reasonably good. However, insertion of an a-i layer even a few nanometers thick at the p-i interface could result in severe performance deterioration of the PIN device. The reflectance spectra of nc-Si:H and a-Si:H devices vary as a function of wavelength and the spectra determine the improvement in the device characteristics, when the incubation layer is eliminated.

5.2.6.1 Short-Current Density, Fill Factor (FF), Open-Circuit Voltage, and Conversion Efficiency of a PIN Solar Cell Using nc-Si:H

It is important to mention that the open-circuit voltage, short-circuit current density, fill factor, and conversion efficiency of the PIN solar cell fabricated using a modified pulsed-PECVD technique vary as a function of temperature during growth of the nc-Si:H i-layer. However, optimum values of these parameters for a given i-layer thickness are observed around 0 ± 25°C [2]. The open-circuit voltage varies approximately from 425 to 475 mV, the short-circuit current density varies from 17 to 23 mA/cm^2, the fill factor varies from 0.63 to 0.68, and the conversion efficiency varies from 6.1 to 7.5 percent over this temperature range during nc-Si:H i-layer growth.

It is important to point out that the short-circuit current in a solar cell is dependent on the temperature of the surface. In the case of crystalline solar cells, short-circuit current increases roughly by 0.05 percent to 0.07 percent per degree Kelvin, whereas in the case of amorphous solar cells, it increases about by 0.2 percent per degree Kelvin. The lower temperature coefficient of amorphous silicon devices is one of the reasons why amorphous silicon devices perform better than crystalline silicon devices [1], particularly in hot summer months. The open-circuit voltage is dependent on the surface ambient temperature and the solar panel installation location. The solar panel surface temperature can reach as high as 40°K over the ambient temperature, which can increase the open-circuit voltage, if the panels are installed outdoors. Note the open-circuit voltage increases very rapidly with the illumination density until it reaches the saturation level. Typical open-circuit voltage varies from 500 to 600 mV for crystalline silicon solar cells and from 600 to 900 mV for amorphous silicon solar cells [1].

The relative quantum efficiency of the nc-Si:H-based PIN devices at optimum temperature varies from 75 percent at a wavelength of 0.4 μm to 95 percent at a wavelength of 0.6 μm to 48 percent at a wavelength of 0.8 μm. The PIN diode quality factor varies as a function of thickness of nc-Si:H material in a PIN device.

Based on this performance data, PIN solar cell designers believe that thicker i-layers ranging from 0.8 to 2.2 μm will provide improved device performance. For example, the diffusion length improved from 0.8 to 1.2 μm when the thickness was increased from 1 to 3 μm. In addition, the diode quality factor increased from 1.4 to 1.8 when the device thickness increased from 1 to 3.2 μm and beyond. All these improvements are due to increased recombination within the bulk material. In other words, the material properties improve with thickness. Improvements in device efficiency are possible with suitable texturing for nc-Si:H film growth and using zinc-oxide/silver (ZnO/Ag) as back contact leads to a nonequilibrium condition, which suppresses the dust, thereby reducing the structural defects in the nc-Si:H films. However, using the modified pulsed-PECVD technique, cell designers can significantly reduce, if not eliminate, the amorphous incubation phase at the p-i interface. In summary, the device efficiency can be improved by reducing the amorphous incubation at the p-i interface by optimizing various deposition techniques. The pulsed-PECVD deposition process indicates the i-layer growth temperature plays a key role in the film quality.

5.3 Performance Capabilities and Structural Details of Solar Cells Employing Exotic Materials

The efficiencies of conventional planar p-n junction silicon solar cells are not able to exceed 16 percent under laboratory conditions, which may further degrade to less than 13 percent. Due to performance capabilities and limitations of conventional silicon solar cells, material scientists in the last two decades have actively engaged in research and development on crystalline silicon, amorphous silicon (a-Si) and nano-crystalline hydrogenated (nc-Si:H) solar cells. Performance capabilities and limitations of a-Si, a-Si:H and nc-Si:H solar cells have been discussed briefly earlier in this chapter.

Now the critical performance parameters such as spectral response, fill factor, conversion efficiencies under various sunlight conditions, and electrical power output of various solar cells using thin films of microcrystalline silicon (mc-Si), cadmium telluride (CdTe), copper indium selenium (CIS), copper indium gallium selenium (CIGS), and organic films will be identified.

5.3.1 Performance Capabilities and Structural Details

The spectral sensitivity of the materials determines the conversion efficiency, short-circuit current and open-circuit voltage as a function of incoming light or radiation

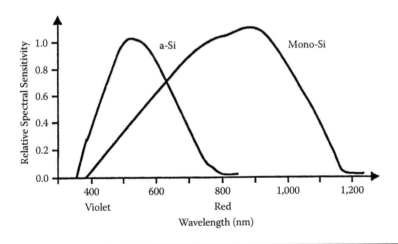

Figure 5.2 Relative spectral response for an amorphous silicon solar cell and a crystalline silicon solar cell.

wavelength. Spectral response levels of amorphous silicon cells and crystalline solar cells (or mono-silicon solar cells) are shown in Figure 5.2.

5.3.1.1 Amorphous Silicon Solar Cell Devices

Open-circuit voltage and short-circuit current variation as a function of incoming solar radiation are shown in Figure 5.3. Device structural details on amorphous silicon solar cells and substrate-type amorphous silicon tandem are illustrated in Figure 5.1. Various material layers, substrates used, and energy band diagrams for these devices are clearly visible in this figure. Note laser sources at appropriate

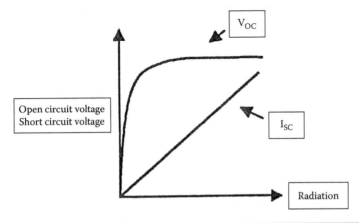

Figure 5.3 Open-circuit voltage and short-circuit current density as a function of radiation intensity.

Table 5.3 Current and Future Efficiency Projections for Various Silicon Solar Cells during the Production Phase (Percent)

Silicon Device Type	Time Frame				
	2000	2005	2010	2015	2020
Amorphous	6	7	8	9	11
Thin-film (crystalline silicon)	13	17	20	22	24
Bulk devices	23	24	25	26	27

wavelengths are widely used in fabrication of solar cells to ensure high production yield, enhanced conversion efficiency and precise device geometrical parameters. For example, the transparent conductive oxide (TCO) layer as shown in Figure 5.1 is deposited on the glass substrate by a laser beam to provide precise layer thickness with excellent fine grain and minimum impurities. Efficiency projections for conventional amorphous silicon, thin-film silicon-based crystalline, and bulk silicon devices are summarized in Table 5.3.

Research scientists believe that optimized cell designs incorporating buffer layers, unique alloys, high electronic quality materials, and appropriate doping gradients will not only enhance conversion efficiency and cell reliability, but also reduce the degradation. Stacked cell and tandem cell configurations (Figure 5.1) will unquestionably improve conversion efficiency, but at higher fabrication costs. At this time, amorphous silicon cells and thin-film crystalline silicon solar cells are widely used in the design of solar power modules, because of the maturity of these materials and deployment of laser-based deposition technologies.

5.3.1.2 Thin Films of Copper Indium Diselenide (CIS) and Copper Indium Diselenide Gallium (CIGS)

Material scientists are deeply involved in improving the performance of CIS solar cells using thin films of this material. Furthermore, solar cell designers believe that thin-film technology must be deployed in the fabrication of solar cells, if high conversion efficiency is the principal requirement, regardless of the semiconductor material involved in the device fabrication. Recent design data on CIS devices [1] indicates that a CIS solar cell has the potential to reach the limiting efficiency of crystalline silicon-based devices. The data further indicates that a CIS device has a much higher limit than an amorphous silicon cell under the same solar illumination intensity and operating environments. Solar cell research scientists predict that the measured efficiencies of CIS devices under laboratory environments would be higher by about 8 to 10 percent over the efficiencies expected during the production phase. Laboratory and production efficiency projections [1] for a-Si, CIS and CIGS cells are summarized in Table 5.4.

Table 5.4 Efficiency Projections for a-Si, CIS and CISG Solar Cells under Various Testing Environments as a Function of Development Time Frame (Percent)

Device type	Time Frame				
	2000	*2005*	*2010*	*2015*	*2020*
Amorphous silicon (production)	6	7	8	9	10
Amorphous silicon (laboratory)	14	15	16	17	18
CIS (production)	8	11	13	14	16
CIS (laboratory)	18	20	21	23	24
CIGS (laboratory)	18	20	22	23	24

Source: Data from Reference 1.

Projected efficiency data for various cells under AM1 condition reveals that the limiting laboratory efficiency for a-Si devices is about 18 percent, will reach close to 28 percent in the year 2080. On the other hand, the CID devices are expected to provide higher efficiency in the future before reaching a limiting laboratory efficiency of 28 percent, which is higher than about 10 percent over the a-Si devices. Note the conversion efficiencies for copper indium diselenide gallium (CIGS) solar cells under laboratory conditions are very close to those of CIS devices. However, the limiting laboratory efficiency for a CIGS cell could reach to roughly 28 percent earlier than a CIS device. It appears that the CIC solar cell technology will play a major role in the design and development of solar power sources.

5.3.1.3 Benefits and Drawbacks of Ternary Compound Semiconductor Material Used in the Fabrication of CIS and CIGS Solar Cells

The projected performance data presented on solar cells incorporating thin films of ternary compound semiconductor materials such as CIS and CIGS seem to indicate significant improvement as far as the conversion efficiency is concerned. Thin-film processing and laser deposition techniques have also made contributions to higher efficiency and reliability. The addition of gallium seems to enhance the cell efficiency. However, the complexity in handling these materials, additional fabrication costs, and their toxicity require further investigation in terms of economics, the environment, and mass production.

Crystalline silicon solar cells are preferred over amorphous silicon devices because of lower conversion efficiencies of the amorphous devices. Sputtering of metal films with subsequent selenization is best suited for thin-film process in CIS devices. Since the expected production phase efficiencies for CIS devices are in

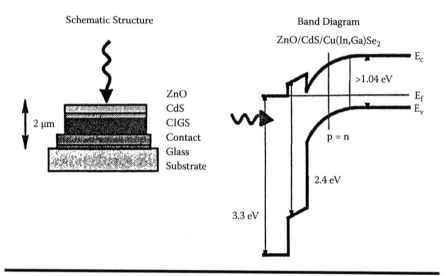

Figure 5.4   Structural details and band diagram of a CIGS solar cell.

excess of 16 percent, the CIS solar cell will be the frontrunner with respect to the conversion efficiency of thin-film solar cells.

The future of CIGS solar cells is dependent on the fabrication processing cost, design complexity, and the severity of phase composition of the ternary semiconductor compound materials involved. Furthermore, the energy levels of this complex material lie in the valence or the conductance band. Note the formation of a copper-depleted defect layer at the surface of the p-type film could slightly degrade the conversion efficiency of CIGS solar cells. Structural details of a CIGS solar cell are shown in Figure 5.4. Research scientists have identified the presence of some toxic substances in CIS and CIGS solar cells and have recommended precautions during the fabrication, testing, and installation phases.

5.3.1.4 Cadmium Telluride (CdTe) Solar Cells

Solar cell scientists have investigated the use of thin-film CdTe material in the design and development of solar devices, but they are not impressed with the performance of these devices. Even after a long design and development duration, the CdTe devices have demonstrated conversion efficiencies below 16 percent and large-area solar module efficiencies hardly close to 10 percent. The initial CdTe devices were designed using CdTe single crystals. RCA developed the first alloy-type pn-junction CdTe cell with conversion efficiency around 2 percent. USSR scientists demonstrated their CdTe cell efficiency better than 4 percent. Later on research scientists designed and fabricated a pn-heterojunction CdTe solar cell as shown in Figure 5.5 using a p-type material in conjunction with n-type CdS material, which demonstrated conversion efficiency better than 6 percent. After this

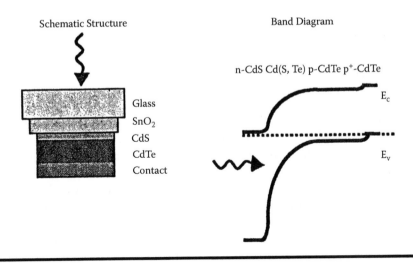

Figure 5.5   Design of a CdTe solar cell and its band diagram.

unique historic development, research scientists noticed serious fabrication and performance problems such as the difficulty of doping p-type CdTe material, inability to achieve low-resistance contacts, and excessive recombination losses associated with the junction interface.

Continuous research and development activities on an n-type CdS/p-type CdTe heterojunction-based CdTe device demonstrated an efficiency close to 10 percent in 1982. Incorporating a CdS thin layer acting as a buffer layer and using a transparent conductive oxide (TCO) coating, the research scientists have pushed the laboratory efficiency close to 16 percent, which would be realized between 12 to 13 percent during the production phase.

5.3.1.4.1 Impact of Material Properties and Film Deposition Techniques on CdTe Device Electrical Performance

Material scientists have found CdTe the most ideal material for incorporation in the design and development of thin-film solar cells due to its several material properties. This material has an optical bandgap close to the optimum for solar energy conversion and is very easy to handle for thin-film deposition using laser technology. Because of the tolerance of this material to defects and grain boundaries, low-cost and simple electro-deposition and screen printing processes are possible. Thin-film growth can be accomplished very close to the equilibrium condition, leading to higher conversion efficiency and lower fabrication costs. Recently researchers have found different techniques for CdTe film deposition, which offer flexibility with respect to large-scale manufacturing and in-line processing with high deposition speed. Note high deposition speed permits handling of large batches of solar cells leading to CdTe development plan scaling best suited for multimegawatt solar power sources.

5.3.1.4.2 Major Technical Issues to be Addressed for CdTe Solar Cells and Modules

Some critical issues on CdTe solar cells and modules must be addressed before mass production using CdTe technology can be embarked upon. For economical production and manufacturing of these devices, some critical issues must be addressed, such as junction formation with zero defect, crystallization of the films with high quality, and formation of stable back contacts with minimum resistive losses [1]. It is important to mention that the fabrication of heterojunction CdTe solar cells is more complex than conventional p-n junction planar devices, because it requires intermixing of the CdS and CdTe layers at a single optimum temperature without degrading the quality of the films or layers. Furthermore, the intermixing produces a graded gap structure as illustrated in Figure 5.5, instead of an abrupt heterojunction structure. As stated earlier, the architecture of this solar cell allows production of solar devices with large areas, leading to production of solar cells with high yield and minimum production costs. It is extremely important to point out that both Cd and Te are toxic materials, which require special handling during fabrication and testing phases. However, the CdTe compound itself is quite stable and harmless, according to the research scientists who are working closely with these materials.

5.3.1.5 Solar Cells Using Thin Films of CdHgTe

In early 2001, the *Journal of Materials Science* reported the performance capabilities of solar cells deploying cadmium mercury telluride (CdHgTe) thin films electrodeposited on ITO glass/cadmium sulfide substrate (ITO/CdS substrate) [3]. Limited performance data such as series resistance (R_s), open-circuit voltage (V_{oc}), ideality factor (A), fill factor, short-circuit current density (J_{sc}), reversed saturation current (I_{sat}), and conversion efficiency as a function of contact processing parameters are available. However, these data do not reveal the current state-of-the-art device parameters. Preliminary performance data available on these solar cells to date indicates that annealing temperature, duration, and quenching cycles have considerable impact on the open-circuit voltage and conversion efficiency of the CdHgTe devices.

5.3.1.5.1 Device Fabrication Processing Steps

The CdHgTe solar cell is a heterojunction solid-state device comprising an n-type CdS film and a p-type CdHgTe film deposited on an ITO-coated glass substrate. The surface of the ITO-coated glass substrate must be absolutely clean using multiple rinsing in boiling water to remove even minimum traces of detergent and finally with an ultrasonic cleaner. After cleaning, the films must be dried in flowing hot air to eliminate moisture from the film surfaces.

The CdS films are first electrodeposited. The CdS films so deposited are highly resistive and must be subjected to rapid thermal annealing in a vacuum to reduce the CdS film resistivity well under 0.15 ohm-cm. After that the electrodeposition of CdHgTe films can be performed. Post-deposition annealing, rinsing, and drying procedures will result in recrystallization of CdHgTe conversion from n-type to p-type, leading to the formation of an n-CdS/p-CdGgTe heterojunction solar cell. Gold-copper (Au-Cu) circular contacts of appropriate dimensions are thermally evaporated on the CdHgTe surface to achieve low contact resistance needed for higher cell efficiency. Finally, a room temperature curing-type silver (Ag) cement must be applied over the evaporated metal film to provide the mechanical stability to the contacts and to enhance reliability of the device. Currently, no data on device reliability and their operating capabilities under harsh outdoor conditions are available. In addition, the conversion efficiency well below 8 percent under laboratory conditions is not very encouraging. The device fabrication procedures are complex and cumbersome. Based on the above statements, deployment of CdHgTe solar cells in solar power modules can be hardly justified. These devices are not very attractive for mass production, unless significant improvement in efficiency and reduction in fabrication costs is forthcoming in the near future through comprehensive research and development activities and backed by extensive laboratory tests and evaluations.

5.3.1.5.2 Critical Performance Parameters of CdHgTe Solar Cells

As mentioned earlier, saturation current density, open-circuit voltage, series resistance, contact resistance, conversion efficiency, fill factor, and the heterojunction diode quality factor (A), also known as ideality factor, are the most desirable performance parameters. Accurate determination of the light intensity-dependent parameters such as diode quality factor and series resistance is difficult in a polycrystalline solar cell such as the CdHgTe device due to uncertainties in the measurements of light data and hence the photogenerated current. Measurements of the series resistance and ideality factor must be obtained under dark and light conditions. The performance parameters can be fairly estimated [3] by analyzing the test results for various Au-Cu contacts processing under the following conditions:

- Device under no contact processing (device A).
- Contacts annealed at 200°C for one minute and quenched (device B).
- Contacts annealed at 150°C for 15 minutes and repeatedly quenched following each minute of annealing (device C).
- Contacts annealed at 150°C for 5 minutes.
- Contacts annealed at 150°C for 5 minutes and repeatedly quenched following each minute of annealing (device E).

Degradation in open-circuit voltage can occur from the diffusion of the Au-Cu contact within the bulk either because of the higher temperature or longer annealing duration. Presence of forward current in the device indicates that the p-n junction is leaky. As far as the series resistance is concerned, its highest value occurs under dark condition (i.e., in the absence of light), while the lowest value occurs under illumination condition. The post-deposition contact annealing will decrease the series resistance and improve the short-circuit current density. However, annealing temperature higher than 150°C and longer annealing duration will degrade the open-circuit voltage and will make the p-n junction leaky. Higher collection efficiency of the device is contingent on the grain size of the films and lattice constant of the CdHgTe films.

The lattice constant of the CdHgTe film (d_{hkl}) can be calculated using the following equation:

$$d_{hkl} = [a/\sqrt{(h^2 + k^2 + l^2)}] \tag{5.1}$$

where a lattice parameter of the film has a value of 0.6473 nm for the unprocessed films or nonannealed films and a value of 0.6474 nm for the annealed films; h, k, and l are the physical dimensions of the film in nanometers. According to the research scientists who have worked on these films, in all cases the calculated value of the lattice parameter a lies between the corresponding standard values of the lattice parameters of the HgTe film as 0.6465 nm and of the CdTe film as 0.6478 nm.

Contact material and deposition technique are critical in the design of contacts. CdS/CdTe-based solar cells continue to suffer from the contact problem because of the high work function, which is 5.76 eV or above [3] and no metal or alloy has a higher work function. One of the most efficient methods used to produce good ohmic contacts on p-type CdTe films involves the electrodeless deposition of a noble metal from a solution. An Au-Cu contact material offers the acceptable electrical performance. Electrodeposition of Au-Cu contacts on the CdHgTe film surface not only yields high conversion efficiency, but also uniform current-voltage performance across to adjacent contacts with minimum costs. However, electrodeless deposition of nickel-platinum (Ni-Pt) onto the surface of the CdTe film offers the highest efficiency and stability for the solar cells. In other words, the electrodeless deposition of ohmic contacts on p-type CdTe film surface using a noble metal such as Ni-Pt yields the highest cell efficiency, if slightly higher costs are acceptable.

Comprehensive review of the published technical papers on the subject indicates that post-deposition, annealing process parameters of Au-Cu contacts, intermittent quenching, and presence of oxygen-rich ambient play critical roles in the development of lower-resistance, high performance contacts. The best performing solar devices are obtained when the Au-Cu contacts are annealed at 150°C for 5 minutes, interlaced with quenching process after each minute of annealing.

Consistency in the long range performance of solar cells in terms of spectral response, quantum efficiency, and stability of the device is dependent on the duration of storage. The performance of most devices or systems deteriorates in terms of efficiency, reliability, and stability after long storage. For solar cell devices, reduction in power output and conversion efficiency is observed due to aging. Solid-state devices without encapsulation will experience more performance deterioration after long storage in a dry environment. However, in the case of well-designed CdS/ CdHgTe, the critical performance parameters, namely, the open-circuit voltage and the short-circuit current density that impact the device efficiency, have not experienced significant losses even after 2 years of aging. This indicates that CdS/ CdHgTe devices enjoy good performance stability even after aging.

These devices have demonstrated acceptable spectral response over 500 to 800 nm wavelength. In addition, a fairly flat response to all photons between 550 to 800 nm has been, which is a major part of the incident solar radiation. Quantum efficiency is the most important performance parameter of a solar cell. Rapid increase in external quantum efficiency for wavelength less than 850 nm and more than 500 is due to fundamental absorption edge of CdS and CdHgTe films. The external quantum efficiency between 500 and 600 nm is significantly higher under zero bias conditions. The quantum efficiency experiences very slight improvement under bias condition. The increase in quantum efficiency under bias conditions is a good indicator for better carrier collection.

Material scientists believe that contact processing schemes described in Section 5.3.1.5.2 have significant impact on the performance parameters of the CdS/ CdHgTe solar cell devices. Five different contact processing schemes have been identified, as devices A, B, C, D, and E. Each scheme has defined the requirements for annealing temperature and quenching duration needed to improve device performance. Performance improvements in various CdS/CdHgTe devices are summarized in Table 5.5.

Table 5.5 Performance Improvements in CdS/CdHgTe Devices Due to Au-Cu Contact Processing Schemes

Parameters	Device A	Device B	Device C	Device D	Device E
Short-circuit current density (mA/cm²)	9.67	25.8	25.6	24	23.5
Open-circuit voltage (mV)	510	422	405	512	538
Fill factor	0.32	0.29	0.40	0.34	0.38
Cell efficiency (%)	1.98	3.92	4.93	5.06	5.84

Comprehensive review of the summarized results indicate that contact annealing at 150°C for 5 minutes and repeated quenching following each minute of quenching offers highest conversion efficiency of this device. Optimization of annealing and quenching requirements prescribed for device E can boost the short-circuit current density to 26.8 mA/cm^2, open-circuit voltage to 560 mV, and conversion efficiency to 7.4 percent under zero bias conditions. The main cause for poor efficiency of these devices is the high resistive losses of the Au-Cu / CdHgTe contacts. If the contacts are made from electrodeless deposition of Ni-Pt Nobel alloy, an efficiency improvement of about 2 percent is possible. Further efficiency improvement close to 2 percent is possible through the structured surface characterized by sharp pyramidal grains. This particular source of improvement is strictly due to multiple surface reflections in the incident solar light. If all these design features are implemented, the CdS/CdHgTe solar cells can yield conversion efficiencies close to 12 percent. Even if further design improvements such as exotic contact configuration and ultralow-loss thin films with enhanced grain structures are incorporated, the maximum conversion efficiency of these solar cells will not exceed 15 percent under laboratory testing environments. Furthermore, fabrication and production costs for these devices with added design features will be higher. Reliable estimates for device production cost, operational reliability, and device stability are currently not available.

5.3.2 MIS Solar Cells

The latest research studies on measuring interface state (MIS) solar cells reveal that efficiencies of 12.5 percent and 8.9 percent have been demonstrated for single-crystal silicon and Wacker polycrystalline silicon solar cells [4]. The efficiency data obtained on cell area of 2 cm^2 and under simulated illumination intensity of 100 mW/cm^2 can be further improved by incorporating high quality thin films and enhancing grain boundary effects. MIS devices using silicon ribbons exhibit lower open-circuit voltage and lower conversion efficiency. The reduction in voltage is strictly due to interface state effects.

Research scientists have designed, developed, and tested MIS devices [4] using polycrystalline, ribbon, and (100) single-crystal silicon substrates with interfacial oxide layer of 40 to 60 Å thickness and incorporating Cr, Al, Ti, and Cu as Schottky metal. Computer simulation is necessary to predict the effects of interface states, interfacial oxide layer thickness, and Schottky metal on the open-circuit voltage. The influence of grain boundaries on interface states and open-circuit voltage must be thoroughly investigated. In addition, comprehensive studies on different insulators and Schottky metals must be undertaken to prove their suitability for the interface state profiles.

Reliable cost data on the fabrication and production of these devices are currently not available, because a limited number of devices were developed just to demonstrate the impact of interface states, interfacial oxide thickness, and Schottky

Table 5.6   Performance Data on MIS Solar Cells Based on Interface State Study Investigation Results

		Device Parameters			
Metal Type	Work Function (eV)	Silicon Type	Diffusion Length (μm)	Time Constant (sec)	V_{oc} (mV)
Cr	4.32	Ribbon (EFG)	23	0.00145	495
Cr	4.32	Ribbon (IBM)	32	0.03400	506
Cr	4.32	Polycrystalline	60	0.03100	550
Cr	4.32	(100) Single-crystal	70	0.07800	584
Ti	4.28	----------DO---------	70	0.04400	545
Al	4.25	----------DO---------	70	0.02400	573
Cu	4.5	----------DO---------	70	0.04400	364

metals on the open-circuit voltage. A few MIS devices were developed also to investigate the effects of substrate choice and insulator characteristics, Schottky metal tunneling effects, recombination time constants, and minority carrier diffusion length. For optimum performance, silicon substrates other than (100) single-crystal must be surface treated prior to fabrication processes. Note the interface states are in equilibrium for semiconductor film thickness greater than 15 Å, according to material scientists.

Computer simulations performed by various research scientists reveal that the lowest interface state density occurs in (100) single-crystal silicon at midgap. Polycrystalline silicon substrates and ribbon silicon show higher interface state densities, which are responsible for lower open-circuit voltages, as shown in Table 5.6. Higher interface state density is reported by various material scientists at the conduction band edge, with a minimum at midgap. As mentioned earlier, chromium-based MIS devices fabricated on (100) single-crystal silicon substrates offer the highest open-circuit voltages. Note excessive interface state density will significantly reduce the open-circuit voltage, when the capture cross-section exceeds 10^{-15} cm^2.

5.3.3 *Schottky-Barrier Solar Cells*

Laboratory investigation on a specially layered Schottky-barrier solar cell (SBSC) with surface area of 1 cm^2 revealed a conversion efficiency better than 9.5 percent, theoretical fill factor 0.67, which will be reduced to 0.42 if the series resistance approaches to 5 ohms. The fill factor of 0.67 is much better than 0.58 for a commercial p-n junction silicon solar cell [5].

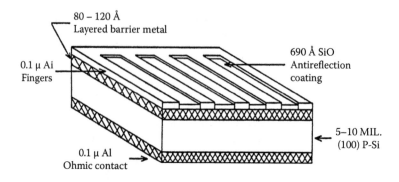

Figure 5.6   Structural details of a layered barrier solar cell showing the ohmic contact, aluminum finger, and antireflection coating.

5.3.3.1 Fabrication Procedure for the SBSC

Structural details of an SBSC device such as the layered barrier metal, antireflection coating, (100) single-crystal p-type silicon layer, and aluminum contact are shown in Figure 5.6. The aluminum ohmic contact must be first formed prior to the deposition of the barrier metal layer. The layered barrier metal consists of a chromium layer of about 50 Å thick adjacent to silicon to provide the proper photovoltaic effect. A copper layer with 50- to 70-Å thickness is deposited over the chromium layer to provide low sheet resistance. A silicon oxide layer of 690 Å (0.069 μm or 69 nm) thick is deposited to produce an excellent antireflection coating for the device. The metal-silicon oxide system on a sliding glass permits light transmission efficiency better than 55.6 percent at a wavelength of 0.6 μm [5].

5.3.3.2 Characteristics of the SBSC Device

The current-voltage (I-V) will show the theoretical effects of the diode quality factor (Q) on the fill factor (FF) and the open-circuit voltage (V_{oc}), while keeping the barrier height parameter (H_B) constant. The open-circuit voltage for this device can be written as

$$V_{oc} = Q\left[(kT/q)\,\ln(I_s/A^*T^2) + H_B/q\right] \tag{5.2}$$

where the quantity (kT/q) is equal to 0.0258 for $T = 300$ K, I_s is the current in the Schottky barrier (typically 30 mA) for a device area A equal to 1 cm², Q is the diode quality factor and is assumed equal to 3, H_B is typically about 0.7 eV, A^* is given as 32 A/cm²-K², and the parameter q is the work function for silicon and is equal to 3.359 eV. Using the given parameters, the parameter A^* is reduced to

$$A^* = [32 \times 1]\ \text{A/cm}^2\text{-K}^2 \tag{5.3}$$

Inserting the assumed values of all the parameters in Equation (5.2), the magnitude of the open-circuit voltage comes to

$$V_{oc} = 3 \ [(0.258) \ \ln (30/32 \times 1000 \times 1 \times 300 \times 300) + 0.7/3.359]$$
$$= 3 \ [(0.258) \ \ln (1/0.95994 \times 10^7) + 0.2084]$$
$$= 3 \ [(0.258) \ (1/16.0773) + 0.2084]$$
$$= 3 \ [0.0016 + 0.2084]$$
$$= [3 \times 0.21]$$
$$= [0.63 \ V) \ \text{or} \ [630 \ \text{mV}]$$

If the diode quality factor is increased to 3.5, the open-circuit voltage increases close to 735 mV. Since the increase in other parameter values has insignificant impact on the open-circuit voltage and conversion efficiency of this device, the diode quality factor must be made as large as possible. Solar cell designers believe that increase in the quality factor can be accomplished by controlling the Schottky metal selection and fabrication procedure. This is possible by incorporating an insulation layer of appropriate thickness between the Schottky metal contact and the semiconductor layer. In addition, the contact resistance must be reduced, if higher open-circuit voltage and enhanced conversion efficiency are the principal design requirements. It is interesting to mention that an SBSC device with junction area of 1 cm^2 demonstrated an efficiency of 9.5 percent compared to 10 percent efficiency for a commercial p-n junction silicon solar cell with area of 20 cm^2 under the same AM1 sunlight illumination. In addition, the SBSC solar cell has a fill factor of 0.60 compared to 0.42 for the same p-n junction cell. It is important to mention that a series resistance of 5 ohms can reduce the fill factor of 0.67 to 0.42. The series resistance can be as high as 5 ohms for SBSC solar cells, whereas it is typically around 1 ohm for well-designed p-n junction silicon solar devices. The fill factors for conventional p-n junction silicon solar cells are higher than that of Schottky-based solar cells. Therefore, higher contact resistance and lower fill factor are responsible for lower open-circuit voltage and lower conversion efficiency in the case of Schottky-based solar cells. Furthermore, the layered concept used in the fabrication of SBSC devices produces high conversion efficiency by allowing independent control of barrier height, optical transmission, and series resistance. Research scientists predict that advances in Schottky-barrier solar cell technology could bring significant improvement in efficiency of such devices in the near future. Since a very small area is needed for fabrication of SBSC devices compared to conventional p-n junction solar cells, there will be a great savings in the fabrication/production costs. However, fabrication costs and reliability data on SBSC devices are not readily available.

5.3.3.3 Dye-Sensitized Solar Cells

Materials scientists have demonstrated that only dye molecules directly attached to the semiconductor surface are capable of injecting charge carriers into the semiconductor material with a quantum yield efficiency exceeding 90 percent. Nanocrystalline dye-sensitized solar devices are based on the mechanism of a fast regenerative photochemical process. Note the dye is the functional element responsible for light absorption and can be separated from the charge carrier transport. In the case of an n-type semiconductor titanium oxide (TiO_2) with a bandgap energy of 3.2 eV, the working cycle commences with the dye excitation due to photon absorption at the TiO_2-electrolyte interface, leading to an electron injection into the titanium oxide layer. The injected electrons can migrate to the front electrode which is a transparent conductive oxide (TCO) layer known as TCO glass and can be extracted as an external current. The dye can be reduced by a redox electrolyte, which also facilitates the charge transport between the counter electrode and dye molecules. The counter electrode must be plated with platinum to provide a low-resistance electron transfer.

Note the light absorption capability of a dye monolayer is relatively small, which limits the photocurrent efficiency with respect to the solar light incident well below 1 percent. Efficiency improvement ranging from 5 to 7 percent can be achieved by treating titanium oxide electrodes with a nanoporous morphology technology. Scientists believe that large-area and low-cost solar cells can be manufactured with minimum complexity. It is clear that the dye-sensitized solar cells will not approach the conversion efficiencies demonstrated by other conventional devices. But the major advantage of the dye-sensitized solar cell is its conduction mechanism based on a majority carrier transport as opposed to the minority carrier transport of conventional p-n junction in organic solar cells. Research scientists predict that even with aggressive research and development activities and large funding allocation, the conversion efficiencies of such devices will not exceed 10 to 12 percent.

5.4 Performance Capabilities of Solar Cells Employing Nanotechnology Concepts

Material scientists have found that it is possible to design photovoltaic solar cells using nanotechnology-based semiconductor materials, including nanowires, nanocrystals, nanorods, and nanodots. Researchers believe that quantum dots have distinct advantages over photosensitive dyes in fabrication of solar cells. Quantum dots have an inherent capability to match the solar spectrum because their absorption spectrum can be tuned with particle size. In addition, quantum dots have demonstrated an ability to generate multiple electron-hole pairs per individual photon, thereby realizing significant efficiency improvement for the photovoltaic solar cells.

5.4.1 Nanowire-Nanocrystal Solar Cells

In May 2007, research scientists at the University of Minnesota developed solar cells using semiconductor nanocrystals known as quantum dots and nanowires. This photovoltaic solar cell was developed with cadmium selenide (CdSe) quantum dots, which were attached to the surface of a single-crystal zinc oxide nanowire [6]. When illuminated with a visible light, the CdSe quantum dots of 3 nm diameter and absorbing light at 550 nm wavelength, injected electrons across their interface with the nanowires. The nanowire lengths varying from 2 to 12 nm provided a direct electrical pathway to the photoanode of the solar cell. This nanowire-nanocrystal solar cell demonstrated short-circuit currents between 1 and 2 mA/cm^2, quantum efficiencies better than 62 percent, and open-circuit voltages ranging from 500 to 600 mV when illuminated with a light intensity of 100 mW/cm^2. However, conversion efficiencies are well below 2 percent. Higher efficiency would require alternate architectures and different nanotechnology-based materials for the nanowires and quantum dots. An optimum solar device would contain quantum dots of variable sizes to absorb maximum energy from the solar spectrum. Despite poor efficiency, quantum dots semiconductor nanocrystals will have a bright future in harvesting the sunlight.

A photon with double the bandgap energy of a quantum dot can produce two electron-hole pairs, while a photon with three times the bandgap energy can produce three electron-hole pairs [6]. This unique response could significantly improve the efficiency of this device in the future. According to research scientists, if the process of multiple electron-hole generation can be fully controlled and optimized, quantum dot devices could achieve efficiencies exceeding 45 percent. However, integrating quantum dots into the solar cell structure is the most basic problem. This involves the larger number of interfaces between the quantum dots and the collecting electrodes interfaces with smooth transfer of the charge. However, scientists believe that to perfect this solar cell technology, it will take at least 5 years before quantum dot solar cells become available for commercial applications.

5.4.2 Solar Cells Using Silicon Nanowires

Research scientists at the Massachusetts Institute of Technology reveal that silicon nanowires could offer some advantages over the more established thin-film technology. Based on numerical simulations, scientists believe that nanowire arrays in some configurations have much lower reflectance compared with thin films, which means that more incoming light is harnessed and converted into usable electrical energy. The researchers further believe that at higher optical energy levels, the nanowires can absorb more incoming sunlight than thin films, thereby significantly improving the conversion efficiency of the solar cell. Typical length of crystalline silicon nanowires varies from 1.2 to 4.7 μm, while the diameter varies between 50 and 80 nm. Such arrays with periodicity of the square lattice of 100 nm, the absorption level varies between 1.1 and 4 eV. Note these nanowires exhibit very little optical absorption capability at lower photon energy levels compared to silicon thin films at the same photon energy levels. However, as the energy frequency increases, the optical absorption of the nanowires rises sharply and exceeds the absorption level of the thin films at 2.5 eV. The absorption level of the silicon nanowires continues to climb, until each of the three lengths reach a plateau at an energy level of 3.0 eV. The absorption of thin films declines gradually after being surpassed by the absorption of the nanowires. Since the silicon nanowires have very low reflectance, antireflection coating may not be required, thereby making the fabrication inexpensive and less complex. The greatest weakness of silicon nanowire technology is that it has poor absorption at low energy levels, which could be overcome with light-trapping techniques or using longer crystalline silicon nanowires [6]. So far no reliable fabrication cost data and conversion efficiency are available for these solar cells.

5.4.3 Solar Cells Using Zinc Oxide Nanorods

Zinc oxide (ZnO) nanorods can play a critical role in the design and development of photovoltaic cells or solar cell devices. Scientists believe [6] that ZnO nanorods are best suited to provide electrical power for battery charged banks that support 12-V emergency lighting and electronic appliances requiring very low electrical power requirements. These solar cells are best suited for small electrical and electronic appliances such as calculators, electronic watches, emergency lighting sources, and other electronic devices requiring the lowest amount of electrical power. Currently, the conversion efficiencies vary from 5 to 8 percent, which are not very impressive. Research scientists predict that optimum configuration of ZnO nanorods, proper selection of nanorod dimensions, and improved fabrication technology could boost the cell efficiency between 8 and 12 percent under laboratory conditions and in the future. Due to poor conversion efficiency and low power capability, these solar cells are not suited for deployment in commercial solar power modules. Furthermore, data on cell reliability and production costs are not available.

5.5 Multijunction Solar Cells

Research scientists believe that multijunction (MJ) solar cells are best suited for applications requiring high solar power levels at enhanced conversion efficiency [6]. MJ cells can be designed in tandem junction configuration or where photovoltaic layers are stacked one upon another. Spectrolab Inc. of Sylmar, California, a subsidiary of Boeing, has demonstrated conversion efficiency close to 41 percent with a multijunction solar cell device. This is the highest efficiency reported on any solar device to date. The MJ solar cell uses the most complicated design configuration, but this solar cell design approach is capable of generating renewable energy at costs that are projected to be competitive with the prices of energy derived from fossil fuels.

5.5.1 Anatomy of a Multijunction Solar Cell

An MJ solar cell consists of three photovoltaic with different bandgap energy levels, as illustrated in Figure 5.7. This solar cell configuration allows more of the available sunlight to be captured and converted into electricity more efficiently. The gallium indium phosphide (GaInP) layer (top layer) is best suited for the short-wavelength regions of the solar spectrum, the gallium indium arsenide (GaInAs) layer (middle layer) is best suited for mid-regions of the spectrum, and the bottom germanium layer is best suited to capture solar energy from the IR regions of the spectrum. Essentially, this particular solar cell is a three-junction "multijunction" solar device. This design offers the breakthrough in the conversion efficiency of the solar cell. MJ solar cells are produced in a metamorphic fashion, also known as the lattice mismatch technique, which allows energy bandgaps to be efficiently manipulated so that the photovoltaic layer or subcell is more efficiently matched to the solar spectrum, thereby realizing significant improvement in collection and conversion efficiencies. The performance of the MJ solar cells can be further enhanced by incorporating built-in miniaturized solar concentrators. The solar concentrators act like magnifying glass elements, which enhance the incoming sunlight with intensity of 240 suns to increase the electrical power from the devices. These solar concentrators are made from inexpensive mirrors and lenses, rather than of more expensive semiconductor materials to reduce fabrication costs.

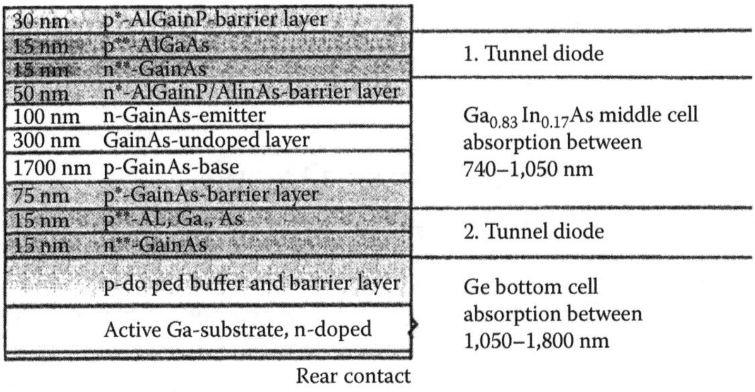

30 nm — p*-AlGaInP-barrier layer	
15 nm — p**-AlGaAs	1. Tunnel diode
15 nm — n**-GaInAs	
50 nm — n*-AlGaInP/AlInAs-barrier layer	
100 nm — n-GaInAs-emitter	Ga$_{0.83}$In$_{0.17}$As middle cell
300 nm — GaInAs-undoped layer	absorption between
1700 nm — p-GaInAs-base	740–1,050 nm
75 nm — p*-GaInAs-barrier layer	
15 nm — p**-Al$_x$Ga$_{1-x}$As	2. Tunnel diode
15 nm — n**-GaInAs	
p-doped buffer and barrier layer	Ge bottom cell
Active Ga-substrate, n-doped	absorption between 1,050–1,800 nm

Rear contact

Figure 5.7 Architecture of a multijunction solar cell consisting of three photovoltaic layers, one stacked over the other, and each layer with different bandgap energy levels to absorb incident energy over various spectral regions.

5.5.2 Space and Commercial Applications

Research scientists involved in the design and development of MJ solar cells predict that a solar panel comprising 1500 cells can deliver electrical power in excess of 30 kW, which is roughly 50 percent more electrical power than could be achieved with silicon concentrator solar cells. MJ solar cell designers claim conversion efficiencies better than 45 percent are possible by year 2010 under laboratory environments, the highest efficiencies ever reported to date for a solar cell device. The electrical power output levels for MJ solar cells are projected better than three to four times compared to conventional single-junction silicon solar devices and are best suited for commercial and spacecraft solar power modules. MJ solar cells incorporating concentrators are more reliable and cost effective compared to conventional silicon solar cells. These devices are highly recommended for terrestrial applications due to their ability to operate with great reliability under space environments.

5.5.3 Market for MJ Solar Devices

Spectrolab Inc. is actively working on the design, development, test, and evaluation of four-junction solar cell design configurations using metamorphic materials with an efficiency design goal of 46 percent under laboratory environments. The company is mass producing and shipping MJ concentrator solar cells with efficiencies close to 36 percent under field conditions. The company claims that this efficiency in the field can be boosted to 42 percent by 2010 [7].

These MJ solar cells have a very bright future in U.S., Japanese, European, and Latin American markets. In a best-case scenario, the Solar America initiative group forecasts a 10 GW solar power installation by 2015 [7], which is expected to reach a level as high as 100 GW, if the exponential growth in solar energy demand continues. It is interesting to mention that solar cell manufacturing factories are currently operating in more than 52 countries. The use of solar energy will increase at a very rapid rate, because the cost of fossil fuels is prohibitive in current world environments.

5.6 Solar Cells Using Polymer Organic Thin-Film Technology

Driven by population growth, global energy consumption, and rapid industrialization in the developing countries, energy planners are compelled to look for electrical energy generation at affordable prices. Today the worldwide electrical consumption is about 13 Terawatt (TW), which will grow to more than 30 TW by the year 2050 [8]. It is important to mention that more than 120,000 TW of the incident solar power is available on the earth's surface. Therefore, design and development of efficient and low-cost solar cells must be accelerated to harvest enough power to meet the world's future electrical energy requirements. The cost of a solar cell is of paramount importance. In addition, the costs of the support mechanical structure needed to support the weight of solar modules and associated electronics must be taken into account in the overall cost for generation of solar energy. High conversion efficiency, improved device reliability, and low installation costs must be given serious consideration during the fabrication, production, and installation phases.

At present conventional crystalline silicon solar cell modules have demonstrated conversion efficiencies close to 13 percent, which translates into an electricity generation rate at 27 cents per unit or kWh, in comparison to 6 cents available from commercial grid electrical sources. Solar energy planners predict the future solar cells must have conversion efficiencies better than 12 percent to be economically viable, if the cost of the solar cell itself approaches zero [8].

5.6.1 Why Organic Thin-Film Solar Cells?

Research scientists at the Stanford Photonics Research Center are pursuing aggressive research and development activities focused on the development of advanced, light-weight, flexible, organic thin-film (OTF) solar cells. These scientists believe [8] that minimum material cost and easy-to-process organic materials are the principal reasons for the development and manufacturing of these solar cells to meet electrical energy requirements at reasonable costs. The low-cost organic material is readily available in sufficient abundance, even to cover very large areas for solar panel installation. The OTF solar cells can be fabricated on light-weight flexible substrates in large areas with low-cost manufacturing techniques, such as roll-to-roll coating, with zero material waste. Currently, the material scientists anticipate no serious reliability and lifetime problems for these devices. If these organic materials can demonstrate lifetimes better than 20 years that is enjoyed by silicon-based solar cells, then one must proceed with the design, development, test, and evaluation of organic thin-film solar cells. Power conversion efficiency of thin-film organic solar cells is not very impressive compared to amorphous silicon devices even after the research and development activities over the last 15 years, according to information published in *Photonics Spectra* (December 2007). Progress on power conversion efficiencies for thin-film organic solar cells is quite evident from the data shown in Table 5.7.

Table 5.7 Progress Made over the Last 20 Years on Power Conversion Efficiencies for Absorption Silicon (a-Si) Solar Cells and Thin-Film Organic Solar Cells (Percent)

	Year				
Solar Cell Type	*1990*	*1995*	*2000*	*2005*	*2010 (Estimated)*
a-Silicon	8.5	10.4	11.8	14.5	18.2
Thin-film organic	0.85	1.2	2.29	5.8	10.0

The efficiency data is at AM1.5 sunlight conditions and under solar radiation intensity of 100 mW/cm^2. Note conversion efficiencies will be much higher under higher radiation intensities or at higher sunlight concentration factors.

5.6.2 Anatomy of the Organic Thin-Film Solar Cell and Its Operating Principle

The OTF solar cell consists of an electron donor organic semiconductor layer and an electron acceptor semiconductor layer located between the top reflecting electrode and bottom transparent electrode. The sunlight is incident on the cell surface through the top transparent electrode. The electron donor semiconductor material must be organic and must have strong absorption bands over the solar spectrum region of interest. Note this is the only layer in this cell that absorbs the sunlight over the spectral region of interest. The electron accepting semiconductor material, known as the acceptor, is usually transparent to the solar radiation and generally consists titanium oxide (TiO_2). The transport electrode is made from indium tin oxide (ITO), while the reflecting electrode is made from aluminum metal. Each solar photon creates an exciton (a mobile way to combine electron and hole in a semiconductor material or in an excited crystal), which migrates into the interface between the acceptor and the donor. The electron and hole separate at the interface, thereby generating an electrical current through the solar cell. The absorption length between the two electrodes varies from 200 to 300 nm. The diffusion length varies from 10 to 20 nm [8]. Note the exciton formed in the donor semiconductor is within the diffusion length from the interface. The OTF device will have a thickness ranging from 250 to 350 nm, which is sufficient for effective sunlight absorption. The designer of this solar cell must have good knowledge in several areas, namely, electrical engineering, materials science, applied physics, and chemical engineering.

In crystalline silicon solar cells, free carriers move easily across the semiconductor layers after they are created by the absorption of photons from the incoming sunlight. However, in organic solar cells, exciton transport occurs by hopping between individual molecules, which is at least four orders of magnitude slower, thereby making the photon absorption more effective. It is important to mention that the principal goal of the organic solar cell is to separate the electron and the hole before the exciton process decays. Since the exciton diffusion length varies between 5 and 20 nm, the donor and the acceptor layers must be structured in such a way that every part of the donor material existing in the light-absorbing spectral region is within 10 to 30 nm of the interface. For optimum cell performance, minimum carrier transportation time and lower recombination losses are highly desirable. Note lower recombination losses require straight pathways in both the donor and acceptor materials, which will ensure that the majority of the excitons are generated within the diffusion length of the interface material layer deployed.

5.6.3 Polymer Semiconductor Solar Cells Incorporating CNT-Based Electrodes

Material types and their properties are critical for the cost-effective manufacturing of the solar cells. Researchers believe that the physics of the solar cell materials and nanostructured materials for electrodes are the key in the development of low-cost organic solar cells. In other words, polymer semiconductor solar cells incorporating carbon nanotubes (CNTs) are considered the most effective light-absorbing electron donors. Note porous TiO_2 is best suited for use as the electron-acceptor material. It is important to mention that polymer materials have direct bandgaps, which interact strongly with the incoming sunlight and exhibit very efficient absorption properties even in extremely thin films with thickness ranging from 100 to 500 nm. This permits the use of very thin films requiring very small amounts of the active materials, thereby realizing reduction in material costs.

The semiconductor polymers offer mechanical integrity and processing advantages necessary for low-cost solar cells. These polymer films can be dissolved in a variety of solvents and can be deposited onto flexible substrates under temperature and pressure best suited for a low-cost wet-processing technique. In addition, scalability to large surface areas with minimum material loss and roll-to-roll coating techniques are possible with polymer semiconductor materials, leading to lower manufacturing costs. These advantages of polymer materials are best suited for highly cost-effective, continuous manufacturing process for low-cost solar cells.

5.6.3.1 Conversion Efficiency of Organic Solar Cells

The conversion efficiency of the organic solar cell is dependent on the I-V characteristics of the organic materials, short-circuit current level, the CNT-based flexible, transparent electrodes, and the dimensional parameters of the titanium oxide electron acceptor. The short-circuit current in solar cells incorporating mesoporous TiO_2 has been observed to be more than three times greater than that generated from solar cells employing flat titanium oxide material. Deployment of CNT-based networks has demonstrated superior performance compared to indium tin oxide (ITO) as an acceptor material. CNT-based electrodes are less expensive compared to ITO. Efficiency data available to date are not very impressive. The most efficient polymer solar cells [8] have achieved conversion efficiencies close to 4.35 percent at the maximum power point and with a fill factor of 0.65. However, research scientists predict that aggressive research and development activities on organic solar cells using improved semiconductor polymers with different bandgaps could achieve efficiencies in the range of 6 to 8 by year 2010 with a distinct possibility of efficiency close to 10 percent in a few years.

5.6.3.2 Organic Solar Cells with Multilayer Configurations

It is important to mention the production of low-cost, high-efficiency solar cells for high power modules will require the optimization of several different parameters involving efficient light-absorbing materials with electron donor properties, low-loss electrode configuration, multilayer device configurations, low-cost substrate processing, and flawless fabrication techniques. In addition, the close collaboration is required between research scientists with a broad range of skills in the fields of synthetic chemistry, chemical engineering, materials science, comprehensive knowledge in organic thin films and optical characterization of materials best suited for photovoltaic cells. The major advantage of low molecular weight electron-acceptor materials compared to large-molecule polymer materials is that they are best suited to achieve amorphous or polycrystalline thin films on flexible polymeric substrates, leading to ultra-low-cost roll-to-roll coating vital to produce solar cells with minimum costs.

Stanford research scientists claim that an organic phase deposition technique will allow deposition of uniform thickness and excellent dopant control, which eliminates the parasitic coating on the chamber walls leading to the fabrication of complex multilayer solar devices with minimum time and cost. The performance of these devices is dependent on different nanostructures, namely, planar ordered nanostructures or disordered nanostructures or buried heterojunction nanostructures. The solar devices with ordered nanostructure have the highest performances. When gold or silver nanoparticles are introduced near the junction regions between the donor and acceptor materials, plasmonic enhancement of small-molecule organic solar cells is observed. The metallic nanoparticles enhance the organic solar cell performance by increasing the optical absorption, leading to the creation of excitons confined to the 10-nm region in the vicinity of the donor-acceptor interface. Increase in optical absorption generates a higher density of excitons in the critical region within the diffusion length of the device junction, leading to higher collection efficiency. Stanford scientists predict that an increase of 50 percent in solar cell efficiency can be achieved by embedding the insulator-coated gold nanorods and nanospheres to a 10-nm region in the vicinity of the donor-acceptor interface. Use of gold nanorods or nanospheres will slightly increase the device fabrication cost.

Some scientists claim that solution-processed random meshes of nanowires can form transparent electrodes with a performance similar to that of indium tin oxide (ITO) films, but it makes the fabrication of large-area solar cells less feasible. Sputtered metal oxide films such as ITO are too expensive for low-cost solar cell manufacturing. Material scientists believe that hybrid transparent-mesh electrodes and combining the metallic nanowires and carbon nanotubes could provide superior electrical performance with significant reduction in device fabrication costs. Improved performance of organic solar devices requires high mobility, low bandgap to match the solar emission spectrum over the range of 1.2 to 1.8 eV, long exciton diffusion length, and high absorption of multiple low-energy photons capable of producing a single higher-energy exciton. These characteristic are best suited to provide the conversion efficiencies of multijunction organic solar cells similar or better than crystalline silicon solar cells with lower fabrication costs.

Development of advanced organic materials with bandgaps at various energy levels that match the solar spectrum components is critical for designing organic multijunction solar cells capable of converting efficiently the solar energy into electricity for home applications. Such an organic multijunction solar cell structure comprises a stack of organic thin films each made of a light-absorbing electron donor material with different bandgaps. The incident sunlight impinges first on the highest bandgap electron-donor material, where most of the energetic solar photons are absorbed and converted into electric current. Note the medium-energy photons and low-energy photons are allowed to pass through without any attenuation to another organic material layer. The remaining incident light impinges on the second layer of medium-bandgap material, where medium-energy photons are absorbed and converted into electric current. Finally, the low-energy photons are absorbed and converted into electric current. Note the higher-energy photons generate a higher output voltage, because they excite a higher-energy exciton and the same current density flows through each of the cell layers, known as subcells. The output voltage of this multijunction organic solar cell is the sum of the voltages generated by each of the subcells. Research scientists believe that the polymer-based, multijunction solar cells using light-absorbing donor materials and incorporating metallic nanostructures to enhance optical absorption near the interface region have the potential to provide conversion efficiencies close to 15 percent by the year 2020. Such multijunction, high-efficiency solar cells offer the least material, fabrication, and installation costs. No reliability or longevity data on these solar devices are available.

5.7 Summary

This chapter identifies potential advanced semiconductor, organic, and nanotechnology materials for the fabrication of solar cells capable of yielding higher conversion efficiencies with lower fabrication costs and higher device yields. Performance capabilities and limitations of solar cells fabricated from cadmium indium selenide (CIS), cadmium indium gallium selenide (CIGS), cadmium telluride (CdTe), and microcrystalline silicon (mc-Si) materials are described. The major emphasis was placed on higher conversion efficiencies with minimum production costs. Performance capabilities and major benefits of amorphous hydrogenated silicon (a-Si:H) and microcrystalline silicon solar cells are discussed. Potential thin-film deposition techniques on specific substrates are described, with emphasis on film quality and uniform surfaces. Critical performance parameters of PIN-solar cells are identified. Advanced contact procedures to achieve low contact resistance and uniform contact surface are discussed briefly. Performance capabilities and limitations of thin-film cadmium mercury telluride (Cd:Hg:Te) solar devices are summarized, with emphasis on conversion efficiency and manufacturing costs. Performance parameters of MIS solar cells using polycrystalline silicon, single-crystal silicon, and silicon ribbons are highlighted, with emphasis on efficiency and device reliability. Impact of Schottky tunneling effects, recombination time constants, and minority carrier diffusion length on the performance of MIS solar cells are discussed in detail. Performance capabilities of Schottky-barrier solar cells are summarized briefly. Conversion efficiency and quantum efficiency of dye-sensitized solar cells are identified, with emphasis on absorption properties of potential dyes. Performance capabilities and limitations of solar cells using nanotechnology-based materials such as silicon nanowires and zinc oxide (ZnO) nanorods are summarized, with emphasis on conversion efficiency and solar incident light absorption capabilities of the materials. Conversion efficiencies, fabrication complexity, fabrication costs, and electrical power generating capabilities of stacked multijunction solar cells and V-groove multijunction solar cells are summarized in detail. Major benefits and potential applications of multijunction solar cells are identified. Performance parameters of polymer solar cells using organic thin films are described. Benefits of incorporating carbon nanotube (CNT) electrodes in polymer solar cells are highlighted. Organic solar cells can be manufactured with minimum costs, but they suffer from very low conversion efficiencies, ranging from 4 to 6 percent. Conventional single-junction silicon solar cells have demonstrated efficiencies close to 16 percent under laboratory conditions, which will decrease to about 13 percent in field environments. Research studies performed by the author reveal that multijunction solar cells incorporating various material layers, stacked one over the other, offer the highest conversion efficiencies, in excess of 40 percent. The top material layer offers maximum absorption capability for the solar photon present in the low-energy spectral region, the next layer offers maximum absorption for the solar photon present in the medium-energy spectral region and the last layer offers maximum absorption for the solar photon present in the infrared-energy spectral region. This way maximum solar energy impinging on the solar array is absorbed in various material layers of the multijunction cell, leading to maximum conversion efficiency. No reliability or production cost data are available to date.

References

1. A. Goetzberger and V.H. Hoffman. *Photovoltaic Solar Energy Generation*. New York: Springer, 2007.
2. U. Das, S. Morrison et al. "Thin-film silicon materials and solar cells grown by pulsed PECVD technique," *IEEE Proceedings, Circuits, Devices and Systems* 150, No. 4 (2003): 282–286.
3. Jaya P. Nair, N.B. Chaure et al. "Non-aqueous electrodeposited CdHgTe films for solar cell applications," *Journal of Materials Science: Materials in Electronics* 12 (2001): 377–386.

4. Jin K. Kim, W.A. Anderson et al. "Relating computer simulation studies with interface state measurements on MIS solar cells," *IEEE Transactions on Electron Devices* ED-26, No. 11 (1979): 1777–1781.

5. Wayne A. Anderson and R.A. Milano. "IV-characteristics for silicon Schottky solar cells," *Proceedings of the IEEE* (1975): 206–208.

6. A.R. Jha. *MEMS and Nanotechnology-Based Sensors and Devices for Communications, Medicine and Aerospace Applications.* Boca Raton, FL: CRC Press, 2008.

7. Editor. "Solar future," *Photonics Spectra*, July 2007, 48–50.

8. Gary C. Bjorklund and Thomas M. Baer. "Organic thin-film solar cell research conducted at Stanford University," *Photonics Spectra*, November 2007, 70–76.

Chapter 6

Solar Cell and Array Designs Best Suited for Space Applications

6.1 Introduction

This chapter describes solar cell and solar array designs best suited for space applications. Design requirements for solar cells and solar arrays capable of powering surveillance, reconnaissance, and communications satellites will be very stringent to meet specified continuous power consumption, reliability, launch economy, and longevity requirements. Over the last three decades or so, silicon p-n junction photovoltaic (PV) cells have been widely deployed to meet the long-duration power supply requirements on space vehicles. Solar cells incorporating silicon technology have provided the primary electrical power for most of the space missions conducted by NASA, the Department of Defense, and COMSAT Corporation. However, since early 2000, multijunction solar cells incorporating three or four semiconductor layers capable of absorbing solar energy levels in various spectral regions are receiving the greatest attention for space applications. These multijunction solar cells offer the highest conversion efficiencies and much higher power output levels compared to conventional silicon p-n junction solar cells. It is important to mention that solar cell technology capable of offering lower fabrication cost and higher conversion efficiencies would find wide-scale terrestrial applications. Off course, their ability to operate reliably under harsh thermal and space radiation environments will be the most critical design specifications. The impact of spacecraft stabilization will determine the overall power consumption requirements. The solar array design configuration will depend on the spacecraft stabilization technique, orbit of operation, and the total electrical power consumption requirements. Space-based radiation damage in solar cells is a matter of great concern and this will be addressed in great detail. Potential solar cell materials belonging to mixed semiconductor groups will be investigated, with emphasis on reduction in cell efficiency and output power level as the solar devices age. Drop in conversion efficiency, open-circuit voltage, and current density as a function of operating days or months will be estimated, for solar cells operating under space radiation and temperature environments. Semiconductor materials are identified that possess optimum values of diffusion length and the minority carrier lifetime, which will yield higher conversion efficiency over a period exceeding 15 to 20 years under space radiation and temperature environments with no impact on the overall cell performance.

6.2 Material Requirements for Solar Cells Used in Space

Comprehensive studies performed by the author on the materials best suited for solar cells intended for use in space indicate that enhanced conversion efficiency, low material cost, and high space radiation resistance are the most essential material requirements. In addition, materials must be capable of absorbing maximum incident solar photon energy in the shortest time. In other words, the optical absorption constant as a function of photon energy must be high enough to allow a larger fraction of the carriers generated by the absorption of the solar photons to be produced within a shorter distance from the surface of the incident on the solar cell. The intensity of the solar light at a distance x below the surface of which a flux I_0 is incident cane be given as

$$I(x) = [I_0\, e^{\alpha x}] \tag{6.1}$$

where α is the optical absorption constant for the material as a function of photon energy $h\upsilon$ (eV).

For optimum solar cell performance, the "active region" length, that is, the sum of the diffusion lengths (L_n and L_p) of the two sides of the junction, must be larger in silicon than in any of the desirable materials such as GaAs or CdTe or CdS or GaP known as heterojunction materials. The minority carrier lifetime (τ) plays a key role in shaping the absorption performance of the solar cell material. The minority carrier lifetime is defined as

$$\tau = [L^2/D] \qquad (6.2)$$

where L is the diffusion length and D is the diffusion constant of the material.

It is important to mention that the maximum conversion efficiency of the solar cell is dependent on the optical absorption constant of the material, the distance from the device surface, energy gap of the semiconductor material, and the minority carrier lifetime. If the semiconductor band structure is such that direct recombination dominates, then the minority carrier lifetime is affected by the thermal equilibrium carrier concentrations and other semiconductor parameters [1].

6.2.1 Why Silicon for Space-Based Solar Cells?

The author has investigated potential materials for fabrication of space-based solar cells, with particular attention to semiconductor energy gap (E_g). The materials investigated include silicon (1.1 and 2.3 eV), gallium arsenide (1.38 eV), indium phosphide (1.27 eV), cadmium telluride (1.50 eV), and gallium phosphide (2.24 eV). The semiconductor energy gap plays a critical role in determining maximum conversion efficiency. Solar cell conversion efficiencies as a function of material energy gap for various semiconductor materials are shown in Figure 6.1.

A market survey of the various semiconductor materials reveals that silicon is the cheapest material and is available in unlimited quantity without any restrictions. Material scientists indicate that silicon amounts for approximately 25 percent of the earth's crust. In addition, silicon processing technology is well matured and the electrical, high frequency, thermal, and mechanical characteristics of silicon are fully known. Silicon cells can withstand operating temperatures as high as 125°C

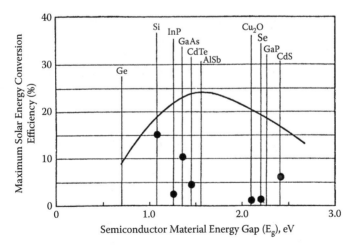

Figure 6.1   Maximum solar energy efficiency for solar cells with various semiconductor materials as a function of energy gap (eV).

with no compromise in electrical performance and reliability. The main reason for selecting silicon is its rapid availability with minimum cost and with no delay in delivering the silicon chips to the manufacturing facility. It is important to mention that silicon n-p solar cells are preferred over p-n solar cells because they are more resistant to space radiation. Furthermore, the trend toward lighter solar arrays for communication satellites and surveillance and reconnaissance spacecraft and higher earth orbits results in wider temperature excursions from sunlight to eclipse periods. These wide temperature excursions and the requirements for long operating life ranging from 15 to 20 years, put great demand on the solar cell design, and the reliability of the intercell and intermodule connections on a solar array.

Material science studies indicate that silicon is the most ideal material for fabrication of space-based solar cells compared to materials such as gallium arsenide (GaAs), cadmium telluride (CdTe), cadmium indium selenide (CIS), cadmium sulfide (CdS), indium phosphide (InP), and cadmium indium gallium selenide (CIGS). Material cost, poor conversion efficiency, and higher fabrication complexity associated with CdS, CIS, and CIGS materials are the principal reasons for not using these materials for space-based solar devices. Note space-qualified CdS solar cells are very costly compared to conventional silicon solar cells. France, Germany, and England have been pursuing research and development activities on CdS, InP, and CdTe solar cells. The studies indicate that the cost per unit area of silicon solar cells decreases as the cell area increases, as illustrated in Figure 6.2. This figure also shows the relative solar cell cost as a function of cell or silicon wafer thickness. Although the conversion efficiencies of GaAs solar cells are higher than those of silicon solar cells, other factors such as density, thermal conductivity, thermal expansion coefficient, and intrinsic carrier concentration, as summarized in Table 6.1, make silicon preferred over the other materials.

The room temperature density of CdTe is 6.062, which is about 2.5 times higher than that of silicon. This means that solar cells and arrays made from silicon will be the lightest compared to other materials, which will not only significantly decrease the launch costs, but will provide enhanced thermal performance, higher mechanical integrity, and improved reliability over a long operating life.

6.2.2 Cadmium Telluride (CdTe) Solar Cells

Extensive research and development activities have been concentrated in the last decade on CdTe solar cells using thin-film technology. Some solar cell companies, including First Solar, Inc., recently have made significant progress in designing thin-film CdTe solar cells with higher conversion efficiencies. Based on continuing test data, the conversion efficiencies for CdTe solar cells have improved from 6 percent in 2005, to 8 percent in 2006, and to 10 percent in early 2008. According to a spokesman from First Solar, Inc., currently these solar devices have demonstrated efficiencies better than 10.7 percent under field operating conditions. Furthermore, the CdTe cell

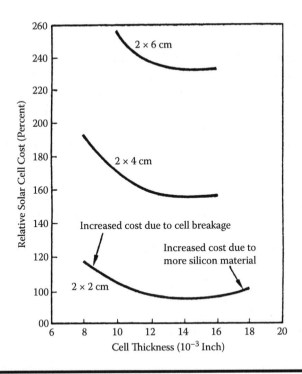

Figure 6.2 **Relative solar cell cost as a function of cell size and cell thickness.**

Table 6.1 **Important Electrical, Thermal, and Mechanical Properties of Si and GaAs Materials at Room Temperature for Solar Cell Applications**

Property	Silicon	Gallium Arsenide
Density (g/cc)	2.329	5.307
Hardness number	950	750
Thermal expansion coefficient (per °C)	2.3	5.7
Thermal conductivity (W/cm ·°C)	1.5	0.38
Intrinsic carrier concentration (per cm³)	1.4×10^{10}	1.4×10^{6}

price is approximately $1.18 per peak watt ($W_p$), which could be about 50 percent cheaper than crystalline silicon solar cells. Because of much lower costs compared to other solar cells, First Solar claims that its production lines are turning out solar modules with power output levels ranging from 10 kW to 200 kW for commercial applications. It appears that CdTe solar cells are most cost effective at present for domestic and commercial applications. Longevity data on CdTe solar cell devices and their ability to operate under space radiation environments are unknown at present.

Following are the unique design features and performance capabilities of the First Solar PV modules:

- First Solar photovoltaic (PV) solar modules are manufactured on high throughput, automated assembly lines that integrate each production step with great precision, from semiconductor thin-film deposition to final assembly and module test in one continuous process. This allows production of solar modules with minimum cost and time.
- Deployment of thin technology produces high energy yield across a wide range of climatic conditions with excellent low-light solar response and temperature response coefficient.
- CdTe solar modules manufactured by First Solar offer high conversion efficiency with minimum cost and complexity.
- First Solar deploys a frameless laminate for their solar muddles, which is robust, most cost effective, and recyclable.
- Solar modules are manufactured in highly automated state-of-the-art facilities to meet stringent quality control requirements and environmental management standards.
- The solar modules manufactured by First Solar are tested by independent, third-party data collection services to verify the authenticity of the measured performance data.
- The operating life of CdTe solar cells is about 25 years, according to First Solar, which is an important factor in selecting the CdTe cells for use in residential and small business applications, where the cost of a solar power module is of critical importance.

6.2.3 Justification for Use of Thin-Film Technology for Solar Cells

Deployment of thin-film technology in fabrication of solar cells permits lower production cost, requires a minimum amount of semiconductor materials, and realizes significant reduction in weight and size of solar cells and modules. In addition, thin-film technology offers higher conversion efficiency and longer operating life for the solar modules. It is important to mention that solar cell cost-per-watt ratio plays a key role in deciding which semiconductor thin-film technology offers the optimum performance, higher reliability, and longer operating life to meet cost-effective design criteria.

Competition with the grid electricity provided by utility companies during peak consumption hours requires the solar cells to cost about U.S. $1 per peak watt, regardless of the semiconductor thin-film technology used in fabrication of the solar cell. This price is called the "grid parity" and it is the holy grail of the photovoltaic industry, which is trying to embark on a multibillion-dollar solar cell program using the thin-film CdTe technology. Several companies are deeply involved in the design and development of CdTe solar cells using a decade-old design based on a thin-film CdTe technology. First Solar has the most advanced manufacturing facility for the CdTe solar cells in the United States. The CdTe cell requires approximately one hundredth of the thickness needed for the silicon solar cell on a low-cost glass substrate, which allows the production of large solar panels with minimum cost and in approximately a tenth of a time it takes for the silicon equivalent. These three distinct advantages of the thin-film CdTe solar cell technology maintain its economic edge over the photovoltaic technologies using thin films of GaAs, CIS, CdS, and CIGS semiconductor materials. In addition, excessive demand for silicon solar cells could lead to a scarcity of the silicon raw material after a couple of decades. It is interesting to mention that prices of silicon solar cells have dropped significantly, which still fluctuate between $3 and $ 4 per peak watt. This price is about three times that of a CdTe solar cell, according to First Solar. Research scientists predict that the annual CdTe solar cell production capacity could exceed 1 gigawatt by 2009. This production capacity could supply one-sixth of that year's estimated global solar cell business, which is currently growing roughly by 50 percent per year.

6.2.4 Performance Capabilities and Limitations of Potential Thin-Film Technologies

Alternate thin-film technologies such as CIS, CIGS, InP, and CdS are available that can be used in fabrication of solar cells. However, cost-effective performance and cost per watt are the principal parameters, which essentially justify the selection of a specific semiconductor thin-film technology for manufacturing of solar cells and panels. Rival thin-film technologies have been receiving very high levels of investment since the silicon shortage could be felt after 2025.

CIGS thin-film technology has been receiving considerable attention, because the solar cell designers are claiming conversion efficiencies close to 20 percent under laboratory conditions. However, quick production of solar panels using the CIGS technology has demonstrated an efficiency of 12 percent only for the solar panels. Due to lack of performance and cost per watt, no one has come up with a full-blown production assembly line and all the venture capital has been soured. The designers further claim that the CIGS thin film can be deposited with minimum effort. However, cost per watt and reliability data are not readily available, which will prevent the full benefits of the CIGS technology from being realized. Such data on CIS-based solar cells are not readily available and, therefore, meaningful assessment of these thin-film technologies cannot be made.

Another thin-film technology, known as amorphous silicon on glass, is making an impact on the solar cell market. These solar cells use tiny quantities of silicon, but their conversion efficiencies of 10 percent or so have not been improved upon. Oerlikon Solar, a Swiss company, is looking into setting up production assembly lines capable of manufacturing power modules with rating from 40 MW to 60 MW per year. Oerlikon is currently producing 85-W solar panels using a 0.3-μm thin-film technology and covering an active area of 1.4 square meters. Since the amorphous silicon film strongly absorbs the visible light; the solar cell output can be increased by 50 percent with addition of a 1.5-μm microcrystalline layer of silicon, which absorbs infrared radiation. However, the all-important cost per watt ratio is not impressive. The company currently estimates a production cost of $1.50 per watt, which could come down to 75 cents per watt by 2010. This production cost forecast appears to be too optimistic. However, a silicon shortage in the near future could pose a serious problem in production and supply of high power solar panels using this thin-film technology.

Multijunction solar cells use thin-film layers of germanium, gallium arsenide, and gallium indium phosphide, which are roughly three times as efficient as the CdTe solar cells manufactured by First Solar. But the production costs are astronomical because the technique requires slow growth rates and the deposition occurs on small germanium substrates. These multijunction solar cells offer conversion efficiencies exceeding 30 percent and are best suited for satellite and terrestrial communication, where high efficiency and reliability are of paramount importance. The high costs can be offset in terrestrial solar cell systems that use large lenses or mirrors to focus sunlight by a factor of several hundreds, which will boost the conversion efficiencies to exceed 42 percent. Due to excessive costs, deployment of such multijunction or multilayer solar cells is limited to terrestrial communications and satellite-based surveillance/reconnaissance applications.

In general, solar cells using thin-film technologies offer lower cost, high production rates, minimum weight and size of the solar modules. Currently, the best conversion efficiency is about 12 percent for thin-film CISG cells, about 10.7 percent for thin-film CdTe solar cells made by First Solar, about 10 percent for thin-film amorphous silicon solar cells and about 35 percent for the high-cost, triple-junction solar cells using three semiconductor layers.

6.3 Performance Parameters for Solar Cells in Space

Conversion efficiency, open-circuit voltage, and current density are the most critical parameters. The values of these parameters are dependent on the materials used in the fabrication of solar cells and their properties. However, the actual magnitude

Table 6.2   Projected Conversion Efficiencies of Solar Cells Using Various Forms of Silicon or Fabrication Formats

Silicon Form	Year			
	2005	2010	2015	2020
Amorphous silicon	14	18	20	22
Crystalline silicon	20	24	26	28
Bulk silicon	21	23	25	27
Thin-film (crystalline silicon)	17	20	23	25

of efficiency and open-circuit voltage is also dependent on the operating environments in space.

6.3.1 Conversion Efficiency of Silicon Solar Cells

Comprehensive studies performed by the author on potential semiconductor materials demonstrate that silicon is the best suited for space-based solar cells and solar arrays. Silicon can be made in various forms. The studies further indicate that currently silicon cells have conversion efficiencies ranging from 15 to 18 percent under laboratory conditions. However, research scientists believe that the projected conversion efficiencies of silicon solar cells will continue to improve at moderate rates. Nevertheless, the conversion efficiency of silicon solar cells will never exceed 28 percent regardless of the silicon technology or form deployed. Projected conversion efficiencies of solar cells using various silicon forms are summarized in Table 6.2.

It is important to point out these conversion efficiencies indicate the efficiencies under laboratory test conditions, which will decrease by 3 to 4 percent due to various losses and adverse space operating environments. Note the field conversion efficiencies will further deteriorate as a function of operating life. Research scientists predict that these conversion efficiencies will deteriorate by 20 to 25 percent over a period of 10 to 12 years, excluding catastrophic failures of some solar cells.

The author is unable to predict design complexity and fabrication processing costs of solar cells made for various silicon forms. Impact of radiation resistance and reliability data as a function of operating life for various space orbits are unknown on these devices. At least one can safely predict the conversion efficiencies of space-based silicon solar cells using various silicon forms.

6.3.2 Relative Solar Cell and Array Costs Using Silicon Technology

Reliable data on silicon solar cell efficiency, operational life in space and space radiation are readily available, because silicon solar cells have been operating in space since March 1958. Several design considerations must be taken into account in the design of solar arrays, such as the radiation effects of solar cells, deterioration in conversion efficiency, and electrical power output level as a function of operating life under space environments, substrate requirements, selection of array voltage, impact of shadows on array performance, and overall solar panel performance in space environments. It is extremely difficult to predict the cost of space-qualified solar cell and array. In 1972, the cost of a silicon solar array varied from $250 to $500 per watt; it is currently under $100 per watt. A typical solar array may contain thousands of silicon solar cells to meet electrical power requirements of several hundreds of watts, which will classify the solar array as the most expensive component of the satellite communications or space-based reconnaissance and surveillance spacecrafts. It is evident from Figure 6.2 that cost per unit area of silicon solar cells decreases as the cell area increases. Figure 6.2 also shows the relative solar cell cost as a function of cell thickness. It is important to note that the cell cost decreases as the cell thickness increases. However, the weight of both the solar cell and array increases with the increase in the cell thickness and the cell area. From a launch cost point of view, light-weight solar arrays are to be preferred. Figure 6.2 identifies the regions of increased cost due to cell breakage and due to more silicon material. Tradeoff studies conducted by the author in terms of critical parameters involved indicate that larger solar cells should be preferred because of cost advantage. The tradeoff studies further indicate that 2 by 4 cm solar cells occupy approximately 20 percent less space per unit area than by the 2 by 4 solar cells, where a large number of cells are required. A solar array comprising 2 by 2 cm silicon solar cells is capable providing electrical power output in excess of 12 kW.

6.3.3 Weight of Solar Cells and Arrays Using Silicon Technology

The weight of the solar cell is dependent on the cell area and the silicon surface thickness. The larger the cell area and thickness the greater will be the weight of the solar cell. The larger the array dimensions the greater will be the array weight and size. The number of solar cells and solar arrays will increase with increase in electrical power consumption. Additional solar cells will be required to meet additional power consumption to compensate for the component losses, reflection losses, attitude stabilization, and control and reduction in cell voltage roughly by 0.5 percent per degree centigrade due to increased temperatures in space. It is important to point out that a solar array with a back surface that can radiate to space will be cooler and, therefore, will have a higher electrical output than an array whose back surface is blocked by the spacecraft or other equipment operating in the vicinity of the spacecraft.

Solar cells are protected by fused silica cover slides for thermal reasons and to enhance resistance to protons and electron radiation encountered in the orbits. The thickness of the fused silica cover slides ranges from 0.004 to 0.020 inches. The cover thickness is selected consistent with radiation protection level required, allowable weight budget and within cell and array costs allowed. Fused silica material are preferred over sapphire and microsheet glass for the cover slides, because of its low cost and excellent stability in space environments. Increase in cost and weight due to cover slides must be taken into account during the weight and cost estimations.

Organic high-polymeric materials must be used as adhesives for bonding the cover slides to the solar cells. Potential adhesives best suited for space applications include epoxies and silicones because of their superior ability to withstand ionizing radiation over long intervals without structural change or change in optical transmittance. Photochemical decomposition results in changes in the optical properties of the adhesives, leading to reduction in the electrical output of the solar cells. By selecting an adhesive that has minimum absorption of the ultraviolet photons and distribution of the energy along the polymer chain, the ultraviolet degradation of optical properties of the adhesive can be minimized. Silicones have this property of minimum absorption and energy distribution and thus, are frequently used for space applications, where operating life exceeding 15 years in space is the principal design requirement.

6.3.4 Maximum Electrical Power Output from Silicon Solar Cells

The principal objective of a solar cell designer is to achieve maximum electrical power output from the space-based solar device with minimum size and weight. Preliminary calculations reveal that power output from a solar cell per unit active area (mW/cm^2) is directly proportional to the cell thickness, but inversely proportional to the resistivity of the material used. Curves shown in Figure 6.3 do confirm these statements. An n-p silicon solar cell with cell thickness of 0.010 inches yields maximum power output of 14.6 mW/cm^2 and 15.5 mW/cm^2 per active area with a silicon chip resistivity of 10 ohms-cm and 2 ohm-cm, respectively. It is clear from these curves that the greater the cell thickness the higher will be electrical power output per unit active area. According to the arguments, solar cells with greater thickness will increase the solar array cost, weight, and size. Note increase in solar array weight and size will significantly increase the launch cost of the spacecraft. It is critical to conduct a tradeoff analysis in terms of solar cell cost, power output per unit active area, and cell size to optimize the cell thickness for cost-effective space-based solar cells and arrays.

6.3.5 Critical Performance Requirements for Solar Arrays for Space Applications

Operation analysis involving material properties, power output per unit active cell area, cell area, cell longevity in space, cell performance degradation, and attitude

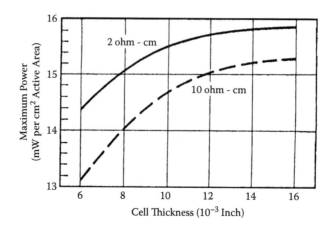

Figure 6.3 **Maximum electrical power output for n-p silicon cells as a function of resistivity and cell thickness.**

stabilization and control must be performed to estimate the power to mass ratio, total array weight, and structural component weight as a function of solar cell efficiency at AM0, that is, under 1 sun condition. Plots shown in Figure 6.4 indicate that cell efficiency plays a key role in determining the total solar array weight, power to mass ratio, structural component weight, and power generating component weight for a 2.5-kW solar array using cadmium sulfide (CdS) thin-film technology [2]. The power-generating weight component includes the weight contributions by the devices, circuits, and electrical accessories exclusively involved in the design of the power-generating subsystem. The structural component weight includes the weight contributions by the support structures, radiation shielding, array cover, and elements needed to maintain high mechanical integrity under space environments. It is important to mention that the total array weight, structural component weight, and power-generating component weight all decrease with increase in solar cell efficiency. However, the power to mass ratio increases rapidly with increase in cell efficiency. Note these curves are only valid at an operating temperature of 55°C. As mentioned previously both the open-circuit voltage and the power output decrease with the increasing temperatures in space. The solar arrays will experience temperature variations from 60°C to 80°C in earth orbit. Higher temperatures exceeding 100°C can be expected in synchronous equatorial orbits.

German and French research scientists reveal that CdS solar cells offer most benefits, except poor conversion efficiencies ranging from 3 to 6 percent. The CdS cells offer 50 percent lower material cost over silicon devices, high stowage efficiency, high array design flexibility, optimum spacecraft configuration at any power level, and 100 times more space radiation resistance over silicon. Major disadvantages include ultra-low efficiency, gradual partial short-circuiting under certain space environments, CdS material not readily available, and about 2.6 times heavier than

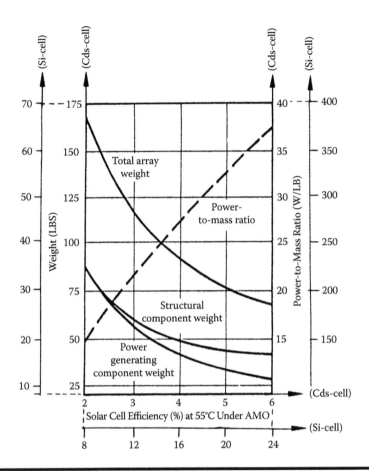

Figure 6.4 2.5-kW solar array weight and power to mass ratio as a function of cell efficiency using CdS and Si devices.

silicon. Because of these major disadvantages, CdS solar cells cannot be justified for space applications, where high conversion efficiency, minimum weight and size, and reliable performance are the principal design requirements.

Since the density of silicon is roughly 2.6 times lower than that of CdS and the silicon is available in abundant quantity compared to CdS, silicon remains the best material for fabrication of solar cells and arrays for space-based solar power generation.

If CdS cells are replaced by silicon solar cells in the 2.5-kW solar array, the values for the total array weight, power to mass ratio, structural component weight, and power-generating component weight will change to new values. The solar cell efficiency at 55°C under AM0 sunlight conditions for both CdS and silicon solar cells are shown on the x-axis in Figure 6.4. These values for various parameters are estimated for the 2.5-kW solar array using cadmium sulfide and silicon. The curves shown for the 2.5-kW solar array can be extrapolated linearly for solar arrays with 10-kW, 50-kW, and 100-kW power output levels. However, the estimated values of these parameters for a silicon solar array with higher power ratings may have ±10 percent errors or higher. The purpose of the curves shown in Figure 6.4 is to provide rough estimates of critical solar array design parameters as a function of electrical power output levels, which will help in the design configurations of solar arrays and spacecrafts.

6.4 Impact of Space Radiation on Solar Cell Performance

Intense space radiation belts exist surrounding the earth consisting of high-energy electrons and protons, which affect the electrical characteristics of the solar cells. Based on Van Allen radiation studies, it is believed that the maximum radiation intensity occurs at an altitude of 10,000 km or 6211 miles. However, a second outer belt also has high intensity at an altitude ranging from 27,000 to 37,000 km. Note the radiation in outer space is characterized by a considerable temporal variation. Space scientists believe there is an omnidirectional flux density of 10,000 protons/ cm^2 with a proton energy in excess of 30 MeV. The best current estimate based on U.S. Air Force studies for the inner radiation belt indicates an electronic flux of about 10^7 particles/cm^2.sec with their energy spectrum extending close to 1 MeV. As regards proton flux, it is found in the energy range of 0.5 keV to 1 MeV; there are roughly 3×10^8 protons/cm^2.sec. The radiation studies further indicate that there are about 10^5 protons/cm^2.sec with proton energy ranging from 1 to 100 MeV.

6.4.1 Performance Degradation from Space Radiation to Solar Cells

The presence of intense radiation belts in the inner belt will reduce the open-circuit voltage and, hence, the power output from the solar cells after about 10^7 seconds (after 3 months) of exposure to the electron flux and after 10^5 seconds (after 30 hours) of exposure to the proton flux of the inner radiation belt. Adverse effects of space radiation will decrease the open-circuit voltage, short-circuit current, cell efficiency, and power output of the device. The higher integrated flux of energetic particles will significantly reduce the short-circuit current more than the open-circuit voltage. Note nuclear radiation can cause changes in various characteristics of a semiconductor material. However, under certain conditions carriers injected by the light will decay exponentially with a time constant, which depends on pre-radiation minority carrier lifetime, capture cross-section of the minority carriers, thermal conductivity, velocity of the carriers, and the integrated flux level. Note any change in the minority carrier life time (τ) will affect the maximum conversion efficiency of the solar cell as well as the diffusion length (L). The initial value of τ is very important in determining the rate of decay of the critical flux φ_c, which is defined as the flux of a given kind of particle that is required to degrade the maximum conversion efficiency (η_{max}) of the solar cell by 25 percent, when the illumination is the solar spectrum on the surface of the earth. Note critical flux is a strong function of the spectral distribution of the source and the spectral response of the solar device. The critical flux is also a function of the initial conversion efficiency of the cell. It is evident that a high efficiency cell will require a smaller flux to reduce its efficiency by 25 percent than a less efficient solar device would need. It is important to mention that even identical solar cells could exhibit wide variations in the critical flux.

6.4.2 Impact of Space Radiation on the Performance of Silicon Solar Cells

For protons in the energy range from 8 to 10 MeV, the value of critical flux lies between 10^{10} and 10^{11} photons/cm^2.sec. A high rate of defect is possible at photon energy (E_p) greater than 200 MeV. However, even at high energy levels, the proton produced damage in space is not significant, because the flux of the protons at the high energy levels is very small. Effects of electron irradiation can be seen in Figure 6.5. This figure provides the estimates for the critical electron flux (electrons/cm^2) as a function of electron energy (E_e) for both silicon n-p type and p-n type solar cells. The difference in damage rates for these two silicon solar devices is clearly visible. In brief, silicon n-p solar cells can withstand higher electron flux with minimum degradation in cell performance.

The effects of electron and proton radiation on solar cell performance were first investigated by RCA Laboratories. Their investigation found that in a given flux n-p solar cells decay by a significantly smaller amount compared to p-n solar cells of comparable efficiency. The investigation further indicated that for protons in the energy range from 8 to 20 MeV, the critical flux for the n-p cells is approximately a factor of 3 to 10 times larger than in p-n silicon solar cells, while for electrons the critical flux is about 30 to 100 times greater for n-p junction silicon cells than for p-n junction silicon cells, as illustrated in Figure 6.5. It is important to point out that variations in radiation damage rates is dependent on the resistivity of the material, differences in the nature of the recombination centers, because protons and electrons produce different kinds of damage on a atomic plane. As stated earlier space radiation has an adverse effect on the open-circuit voltage, short-circuit current, and the conversion efficiency as a function of critical flux levels in the protons and electrons. Solar cells must be designed with high open-circuit voltage to achieve maximum electrical power output levels from the solar cells, which is possible with low values of resistivity, as illustrated in Figure 6.6. Open-circuit voltages are shown in Figure 6.6 for the first-generation silicon solar cells used from 1960 to

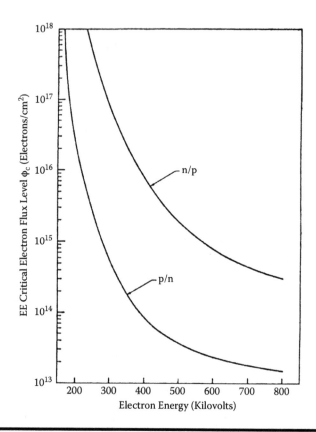

Figure 6.5 **Critical electron flux as a function of electron energy for silicon n-p and p-n solar cells.**

1970 and for the second-generation silicon solar cells designed from 1970 to 1990 as a function of resistivity and operating temperature. Note the open-circuit voltage decreases as the operating temperature increases regardless of the material resistivity. However, reduction of space radiation damage can be minimized by the drift fields in the base region of the solar cell. In other words, the rate of performance degradation due to nuclear particle radiation in silicon solar cells can be reduced by introducing the drift fields into the base region of the solar device. Various degrees of radiation damage are experienced by the base region due to impurity densities in the region, differences of electrostatic potential across the base region, and the starting minority carrier lifetimes. Note the minority carrier lifetime typically varies from 25 nsec for a drift field-free case to 60 nsec for the 1.5 µm junction depth with a drift field of 2000 V/cm. A minority carrier lifetime as low as 0.1 nsec has been reported for a solar cell with a junction depth of 2.8 µm. This indicates that the actual minority carrier lifetimes in a diffused junction are dependent on the minority carrier density and thus vary with the distance below the surface.

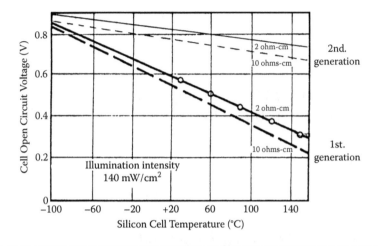

Figure 6.6 **Open-circuit voltage for silicon first- and second-generation solar cells as a function of resistivity and operating temperature.**

6.4.3 Impact of Space Radiation on the Performance of GaAs Solar Cells

Gallium arsenide (GaAs) solar cells offer higher conversion efficiencies due to rapid rise of the absorption constant as a function of wavelength in the GaAs semiconductor material.

Analysis of the spectral response of conventional GaAs solar cells reveals that they yield high efficiency in spite of extremely low values of diffusion length on the order of a few microns. Their high efficiency is due strictly to rapid rise of absorption constant in the GaAs material. Because of this, the maximum efficiency (η_{max}) of a GaAs cell would not change until the critical flux φ_c exceeds that of a silicon cell with comparable pre-irradiation maximum efficiency, because it would take a much larger integrated flux to cause any detectable change in the initial low value of the diffusion length. Based on these arguments, GaAs cells when exposed to irradiation with 17 to 90 MeV protons could tolerate three to five times more radiation than n-p silicon solar cells of comparable pre-irradiation efficiency. The ability to tolerate Van Allen radiation is an important factor in the selection of solar cells for use on spacecrafts or satellites. This critical requirement must be met by the solar cells regardless of the semiconductor materials used in the fabrication of the solar devices. In addition, the solar cell designers must know that actual energy conversion occurs in a relatively small volume of the crystal, which means about 70 μm below the illuminated surface in cadmium sulfide (CdS), about 50 μm below the illuminated surface in silicon (Si) and about 5 μm below the illuminated surface in gallium arsenide (GaAs). Based on this, n-p junction silicon solar cells can tolerate greater radiation integrated flux than p-n junction silicon solar cells. GaAs solar cells appear to tolerate even more radiation resistance from the effects of proton and electron irradiation.

6.5 Effects of Operating Temperature on Open-Circuit Voltage

Studies performed by the author regarding the effects of operating temperatures on the open-circuit voltage and, hence, on the cell electrical power output indicate that both these quantities will decrease as the operating temperature in space increases, whether the cells are fabricated with silicon or with gallium arsenide.

6.5.1 Impact of Operating Temperature on Open-Circuit Voltage of Silicon Solar Cells

The impact of temperature on the open-circuit voltage for the first generation of silicon solar cells was very severe, as shown in Table 6.3. However, advanced research in material science and improvement in device fabrication have cut down considerably on the reduction in open-circuit voltage with increase in operating voltage, as indicated by the curves shown for the second-generation silicon solar cells in Figure 6.6. Reduction in open-circuit voltage for the first-generation silicon solar cells as a function of operating temperature and resistivity is evident from the data shown in Table 6.3.

Performance analysis on silicon solar cell devices indicate that because of reduction in open-circuit voltage the available electrical power output from silicon solar cells decreases approximately by 0.5 percent per degree centigrade rise in temperature. This means that a solar array whose back surface can radiate to space will be cooler and, thus, will have higher electrical power output than an array with a back

Table 6.3 Reduction in Open-Circuit Voltage of the First-Generation Silicon Solar Cells as a Function of Temperature and Material Resistivity

Temperature ($^{\circ}$C/K)	Silicon Resistivity (ohm-cm)	
	2	10
27/300	0.58	0.54
77/350	0.48	0.43
127/400	0.39	0.28

surface blocked by the spacecraft or other equipment operating in the vicinity of the array. However, the latest material research and advanced fabrication processing have shown small reduction in open-circuit voltage and cell power output as a function of temperature and material resistivity for the second generation of silicon solar cells.

6.5.1.1 Low-Energy Proton Damage in Ion-Implanted and Diffused Silicon Solar Cells

It is important to point out that silicon solar cells are vulnerable to low-energy proton damage in space environments. Particularly, unshielded silicon solar cells will experience severe degradation in maximum power output and open-circuit voltage from the low-energy irradiation in space. Rapid performance degradation can be expected from the low-energy portion of the environment in the earth's synchronous orbit. However, ion-implanted silicon solar cells appear to be very much susceptible to the low-energy proton edge effect that diffused n-p junction silicon solar cells. Laboratory tests performed by research scientists indicate that the diffused junction n-p silicon solar cells undergo sudden decrease in output at a critical proton flux of 3×10^{11} protons/cm^2, in addition to an increase in the series resistance of the cell. In the case of ion-implanted solar cells, gradual decline in the maximum output is observed. Research scientists believe that ion-implanted silicon solar cells exhibit superior radiation resistance at both low-energy and high-energy radiation levels. Scientists further believe that the energy levels of the traps introduced in the diffused and ion-implanted n-p junction solar cells differ in their location relative to the Fermi energy level.

6.5.2 Impact of Operating Temperature on the Performance of Heterojunction Gallium Arsenide (AlGaAs-GaAs) Solar Cells

Studies performed by the author on conventional GaAs and heterojunction AlGaAs-GaAs solar cell devices indicate that the absorption constant is very high for the GaAs semiconductor material. The minority carriers are, therefore, photogenerated close to the surface. Consequently, the top layer beneath the surface contributes most of the short-circuit current, which will contribute to both higher conversion efficiency and higher open-circuit voltage. The junction depth in GaAs solar devices is of critical importance. Note the short-current density is a function of junction depth. Furthermore, the short-current density is greatly reduced when the electron diffusion length is shortened to a value comparable to the junction depth. This observation is very important as far as the resistance to radiation damage is concerned. Since the radiation damage leads to shorter electron

Table 6.4 Conversion Efficiency of Heterojunction GaAs Solar Cells as a Function of Operating Temperature and Sun Concentration Factor (Percent)

Temperature (°C/K)	Sun Concentration Factor		
	1 Sun	10 Sun	100 Sun
27/300	20.3	22.4	24.2
77/350	18.2	20.5	22.2
127/400	15.8	18.5	20.6

diffusion length, junction depths shorter than 0.5 μm should be avoided. Note the heterojunction GaAs solar cell essentially consists of a thin AlGaAs layer with thickness less than 0.5 μm and a diffused electrical junction with depth less than 0.5 μm. The former ensures low optical absorption losses and minimum surface-recombination characteristic of GaAs surfaces and the latter ensures increased space radiation resistance [4].

The studies further indicate that both the conversion efficiency and the open-circuit voltage for the heterojunction GaAs solar cell decrease as a function of operating temperature, regardless of the sun concentration factors. The reduction in conversion efficiency and open-circuit voltage as a function of operating temperatures and sun concentration factor (SCF) is quite evident from the data summarized in Table 6.4. The data clearly indicates that the conversion efficiency of the heterojunction GaAs solar devices decreases as the operating temperature increases.

As stated earlier, gallium arsenide is the only semiconductor material when used in the fabrication of solar cells that offers efficiencies exceeding those of silicon devices in spite of extremely low values of diffusion length [4]. The open-circuit voltage also decreases as the environmental temperature increases, regardless of the sun concentration factor. The reduction in open-circuit voltage with the increase in operating temperature is evident from the data summarized in Table 6.5.

Table 6.5 Reduction in Open-Circuit Voltage for Heterojunction GaAs Solar Cells as a Function of Operating Temperature

Temperature (°C/K)	Sun Concentration Factor		
	1 Sun	10 Sun	100 Sun
27/300	1.02	1.08	1.15
77/350	0.92	1.01	1.06
127/400	0.82	0.91	0.98

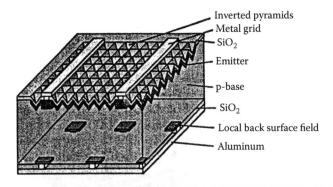

Inverted pyramids
Metal grid
SiO_2
Emitter
p-base
SiO_2
Local back surface field
Aluminum

Figure 6.7   Structural details of a high-efficiency monocrystalline solar cell with conversion efficiency ranging from 22 to 24 percent.

6.5.3 Advanced High-Efficiency Silicon Solar Cells

German research scientists [5] have identified a number of loss mechanisms that can be minimized using advanced fabrication processing techniques, low-resistance contact materials, most efficient antireflection coating, highly doped regions below the contacts, reducing the recombination loss, and surface texturing.

Several high-efficiency solar cells have been described in Reference 1. Particularly, the high-efficiency monocrystalline solar cell appears attractive for space applications. Structural details of this solar cell are shown in Figure 6.7. All three regions of the high-efficiency solar cell as illustrated in this figure, namely, the emitter layer, the base region, and the space charge region between the emitter and base, contribute to low recombination loss. Optimum design features are incorporated in this advanced high-efficiency solar cell, have boosted the conversion efficiency of this cell in the 23 to 24 percent range. The most important design aspects include textured front end with inverted pyramid structures and antireflection coating to reduce reflection losses to a minimum, use of narrow metal contacts from fingers to highly doped regions to minimize recombination losses, deployment of low doping at the surface to achieve good surface passivation, and use of back surface contact with good reflecting properties to reflect light that penetrates to the back of the solar cell.

6.5.4 High-Efficiency Triple-Layer Amorphous Solar Cell for Space Applications

The high-efficiency triple-layer amorphous solar cell, as illustrated in Figure 6.8, consists of three semiconductor layers stacked one over the other and each semiconductor layer is optimized for maximum absorption of the incident solar energy in three distinct spectral regions of the solar spectrum, leading to higher overall conversion efficiency of the cell. The black reflector layer further reduces the reflection

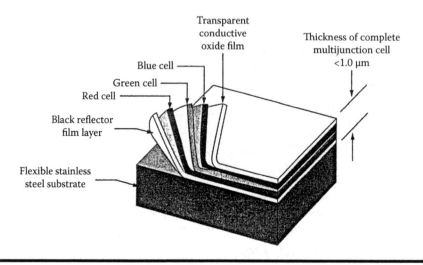

Figure 6.8 A multijunction, triple-layer, amorphous high-efficiency solar cell with conversion efficiency ranging from 21 to 25 percent at room temperature under laboratory conditions.

losses. The transparent conductive oxide film provides protection to the solar cells from dust, rain, and wind. A radiation shield can be installed over the solar array to minimize the damage to the solar cells from low-energy electron and high-energy proton radiation effects. However, this will increase the weight, size, and cost of the solar array. Projected conversion efficiencies for this type of solar cell vary from 21 to 25 percent under laboratory environments. Cost and reliability data are not readily available.

6.5.5 Effects of Proton Energy and Nuclear Particle Radiation on the Performance of Silicon Solar Cells

It is important to mention that performance degradation of the silicon solar cells from low-energy proton radiation can be minimized using effective radiation shielding. Both the collection efficiency and the conversion efficiency can be affected by the space radiation associated with the Van Allen belt. Silicon solar cells with higher base resistivities, mobilities, and starting minority carrier lifetimes generally display better cell performance after radiation damage than comparable solar cells of lower resistivity. Best radiation resistance can be achieved at thicknesses of the drift field layers close to 10 μm. It is worthwhile to mention that the drift field layers greatly influence the short-circuit current degradation and consequently, the cell power output, so that the solar cells can be better designed for improved performance at either low, intermediate, or high radiation doses.

Performance degradation of the solar cell can be very severe under intense nuclear particle radiation depending on the strength of the gamma dose and neutron flux density. A large gamma dose can result in sudden and permanent damage to the solar cell, which can instantly cut the power output of the cell. Protection against nuclear pulse radiation and large photocurrents can be achieved by hardening the base of the device, which will make fabrication very costly. The primary photocurrent can be given as

$$I_{pp} = [50\ (A)\ (d\gamma/dt)] \tag{6.3}$$

where I_{pp} is the primary photocurrent, A is the device junction area (cm^2), and $d\gamma/dt$ is the gamma dose rate (rad/sec), which could vary from 10^6 to 10^8 rads/sec depending on the intensity of the nuclear explosion. Assuming a cell area of 2 cm × 2 cm and gamma dose rate of 10^8 rads/sec, the magnitude of the primary current under these assumptions could be as high as 2 MA. However, the duration of this current level could a few microseconds, which may not be catastrophic to the solar device if tit is radiation hardened.

When a silicon semiconductor material is subjected to neutron radiation, atomic displacement effects occur within the crystal lattice, leading to a significant reduction in both the conversion efficiency and the power output of the solar device. Studies carried out by the author on radiation effects on the performance of solid-state devices indicate that the base width of the cell is the key factor in determining the susceptibility of the solar cell to neutron degradation in nuclear environments. Solar cell performance degradation is a function of diffusion constants and lifetimes of the holes and neutrons, device base dimensions, and the gamma dose rate. The excess current produced by the radiation is known as primary photocurrent, and its initial magnitude can be computed using Equation (6.1). In brief, the ionization effects and photocurrents produced by the gamma dose rates are the principal factors that can create partial damage or catastrophic failure in the solar cells in nuclear environments in space.

Scientists believe that incorporation of drift fields into the base region of the silicon solar cells can significantly extend their useful life in nuclear particle radiation environments [2]. They further believe that optimal drift field configurations are best suited for the achievement of reduced degradation rates due to nuclear particle radiation. Minority carriers generated by electron-hole pair generation can degrade the performance of solar cells due to photon absorption in intense photon radiation fields, but it will not lead to catastrophic failure in the solar device.

6.6 Multijunction Solar Cells for Space Applications

Research and development activities on high-efficiency, radiation resistant, multijunction (MJ) solar cells have been supported since 1998 by the U.S. Air Force, Philips, Wright Patterson Air Force Base, and NASA Lewis Research Center. Recent progress in the characterization, design analysis, and development of high-efficiency, radiation resistant $Ga_{0.5} In_{0.5}P/GaAs/Ge$ dual-junction (DJ) and triple-junction (TJ) solar cells for both space and terrestrial applications is described eloquently in Reference 6. The DJ solar cell is grown on low-cost, high strength, n-type germanium (Ge) substrate, while the TJ solar cell is grown on p-type germanium substrate, as shown in Figure 6.8. These multijunction, high-efficiency, low-cost solar cells are being produced at Spectrolab Inc., a subsidiary of Boeing Company in Pennsylvania. The germanium substrate offers both lower material cost and lower resistivity compared to silicon. The dual-junction low-cost GaInP/GaAs solar cells have demonstrated room temperature conversion efficiencies close to 27 percent at AM0 for space applications. Both the DJ and TJ solar cells have demonstrated final to initial power ratios better than 0.83 after exposure to irradiation with an electronic flux of 10^{15} electrons/cm^2 at 1 MeV radiation level.

6.6.1 Unique Design and Performance Parameters of Multijunction GaInP/GaAs/Ge Solar Cells

This high-efficiency, multijunction GaInP/GaAS/Ge cell with heterostructure configuration consists of three subcells, namely, the GaInP subcell, the GaAs subcell, and the Ge subcell. Each subcell has its own unique material characteristics, which contribute to the overall external quantum efficiency and conversion efficiency of the entire solar cell. The subcell structures are grown by the MOVPE technology with precise control over the layer thickness, composition, and doping levels of the films. Significant performance improvements are achieved with heterojunction structures incorporating various semiconductor layers and interfaces of appropriate materials and layer thicknesses. Improvement in conversion efficiency is strictly due to higher open-circuit voltage and short-circuit current, improved fill factor, reduced reflection and recombination losses, enhanced crystal quality of the epitaxial layers, and optimization of design parameters for the GaInp and GaAs subcells. Electrical power-generating capability and conversion efficiency of the space-based solar cells at the end of life (EOL) are dependent on the cell area, doping levels, subcell design parameters, antireflection coated radiation shield thickness, and power-generating capability and conversion efficiency at the beginning of life (BOL). The bottom germanium subcell (Figure 6.8) in the GaInP/GaAs/Ge multijunction solar cell increases the overall voltage output of the tandem cell, because it is in series with the top two subcells, namely, the GaInP and GaAs subcells. Most of the photons in the incident solar spectrum with energy greater than 1.424 eV bandgap of the GaAs layer are absorbed by GaInP and GaAs subcells. However, the germanium layer still has ample photogenerated current in the 0.67-eV bandgap region of the material, nevertheless, the Ge subcell does not limit the multijunction current capability.

Germanium semiconductor is used as a substrate for both the DJ and TJ solar cells because of minimum material cost and surface hardness. These two properties

Table 6.6   Important Properties of Substrate Materials Best Suited for High-Efficiency, Low-Cost Solar Cells Used in Space

Properties	Germanium	Silicon
Hardness (relative/Knoop)	6.2/650	6.5/950
Melting point (°C)	937	1410
Density (g/cc)	5.36	2.33
Thermal conductivity (W/cm·°C)	0.59	1.47
Bandgap (eV)	0.66	1.11
Resistivity (ohm-cm)[a]	0.01–100	0.001–10,000
Thermal expansion coefficient (%)	5.7	2.3
Relative cost (%)	85	100

[a] Resistivity is a function of impurity.

are considered most vital for high-efficiency, low-cost, space-based solar cells. Note low fabrication cost, high conversion efficiency and surface hardness are the principal design requirements for the electrical power generation equipment aboard communications satellites or surveillance spacecrafts. Important properties of germanium and silicon widely used as substrate materials are summarized in Table 6.6.

Note higher material hardness, lower resistivity, and lower material cost compared to silicon are the principal reasons for selecting the germanium semiconductor material in fabrication of high-efficiency, low-cost space-based solar cells.

6.6.2 *Impact of Temperature in Space on the Conversion Efficiencies of Multijunction GaInP/GaAs/Ge Solar Cells*

Conversion efficiencies of solar cells in space are degraded because of space radiation, space temperature environments, operating life, and shadows/eclipses to which the cells are subjected. The reduction in conversion efficiency due to shadow effects or eclipse is for a very short time, but the reduction in efficiency due to other factors is permanent. However, studies performed by the author on space radiation damage to solar cells indicate that some material such as GaAs will see less damage from the protons, whose energy is high enough (greater than 1 MeV) to prevent them from being arrested close to the active part of the GaAs solar cell. The most damaging low-energy protons below 1 MeV can be effectively shielded by the normal cover glass protection. Reduction in the conversion efficiencies of DJ and TJ GaInP/GaAs/Ge solar cells in space is shown in Figure 6.9. The effect of space temperature and radiation is evident in Table 6.7.

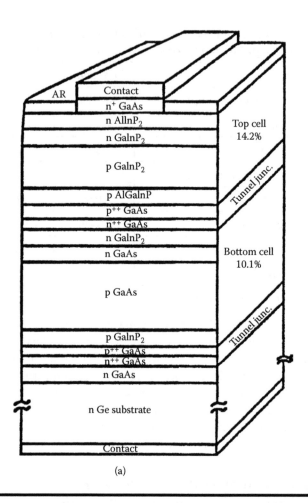

Figure 6.9 Cross-sectional schemes of multijunction cells developed by Spectrolab for (a) dual-junction and (b) triple junction solar cells. *Continued*

6.6.3 Comparison of BOL and EOL Efficiencies of Various High-Efficiency Solar Cells

Conversion efficiency can be optimized using thin-film technology, semiconductor material with unique properties, high-performance antireflection coating, a high-absorbing layer to absorb the incident solar energy, and low-low contact and recombination losses. Performance of both high-efficiency silicon and GaInP/GaAs/Ge solar cells can be optimized by design techniques as mentioned above. Estimates of conversion efficiencies of various optimized versions of solar cells are summarized in Table 6.8. The ratio of power at EOL to power at BOL indicates the reduction in the output of the solar power generating system.

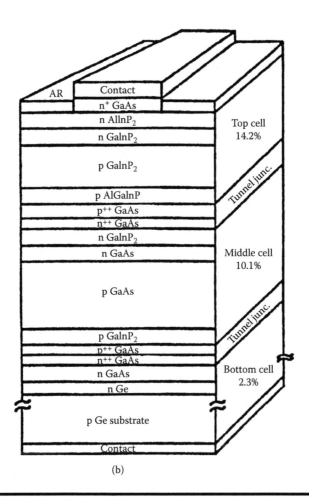

AR
Contact
n^+ GaAs
n AllnP$_2$
n GalnP$_2$
Top cell
14.2%
p GalnP$_2$
p AlGalnP
Tunnel junc.
p^{++} GaAs
n^{++} GaAs
n GalnP$_2$
n GaAs
Middle cell
10.1%
p GaAs
p GalnP$_2$
Tunnel junc.
p^{++} GaAs
n^{++} GaAs
n GaAs
Bottom cell
2.3%
n Ge
p Ge substrate
Contact

(b)

Figure 6.9 *Continued.*

6.6.4 Impact of Space Radiation on the GaAs Subcell

Since the multijunction solar cells shown in Figure 6.9 contain p-type and n-type GaAs layers, the impact of space radiation would not be so serious. This is because the GaAs semiconductor material is a direct bandgap material and has a larger optical absorption coefficient than silicon (an indirect bandgap material). Furthermore, the photovoltaic active region of a GaAs subcell or cell is therefore limited to a relatively shallow depth close to the cell surface. Note the radiation damage beyond this shallow depth, which is generally on the order of 1 μm or so, has a negligible effect on cell performance in terms of open-circuit voltage or conversion efficiency or power output of the solar cell. This means that the minority carrier diffusion length required for reasonably good performance of the GaAs cells must be shorter

Table 6.7 Estimated Values of Conversion Efficiencies of GaInP/GaAs/Ge Multijunction Solar Cells in Space as Affected by Space Temperature and Radiation (Percent)

Temperature (°C)	Duel-Junction Cell (DJ)		Triple-Junction Cell (TJ)	
	BOL-η	EOL-η	BOL-η	EOL-η
40	18.2	16.1	19.4	17.2
50	17.4	15.7	19.0	16.4
60	17.2	15.4	18.6	15.7
70	16.8	15.2	18.0	15.4
80	16.1	14.7	16.8	14.9

Note: The BOL-η and EOL-η indicate the conversion efficiency at the beginning of life and end of life of the solar cell, respectively, after the solar cell was exposed to an electron flux of 10^{15} electrons/cm^2 at 1 MeV radiation level.

Table 6.8 BOL and EOL Efficiencies of Solar Cells with Optimum Designs

Cell Type	BOL-η	EOL-η	P/P$_o$
High-efficiency Si cell	17.4	12.7	0.70
Optimized DJ cell	21.4	18.4	0.86
Optimized TJ cell	23.2	19.9	0.86

than silicon cells before the cell performance deterioration becomes unacceptable in environments involving radiation from high-energy electrons, protons, and neutrons. Both the junction depth and AlGaAs window thickness must be significantly larger than the optimum values desirable for high conversion efficiency and for maximum radiation hardness. Reduction in maximum power output and power ratio in a GaAs subcell as a function of 1 MeV electron irradiation fluence level is evident from the data summarized in Table 6.9. The power ratio is the ratio of the cell power level at EOL to the power at BOL. EOL power indicates the power of the cell after exposure to radiation in space of a specific intensity.

Examination of the data indicates that deterioration in both the maximum power output and power ratio is very serious as the radiation approaches an electron fluence level or density level of 10^{15} electrons/cm^2 at 1 MeV space radiation intensity.

When solar modules are installed to power the communication circuits and devices aboard communications satellites, the designer must upgrade the solar power module output so that all sensors, circuits, and devices will continue to

Table 6.9   Reduction in Maximum Power Output and Power Ratio in a GaAs Subcell as a Function of Electron Fluence Level (electrons/cm²; MW)

Electron Radiation Fluence (electrons/cm²)	Power Output	Power Ratio
10^{12}	86	1.0
10^{13}	83	0.97
10^{14}	80	0.91
10^{15}	66	0.77
10^{16}	44	0.52

operate over the stated life of the satellite. Upgrading of the solar power generating system is essential due to degradation of solar cell performance by Van Allen radiation. Degradation in solar cell performance includes drops in open-circuit voltage, conversion efficiency, and power output. Since most of communications satellites operate in the near earth radiation region, the impact of 1 MeV electron radiation is of immediate concern and, thus, the solar cells and the solar panels must be hardened to this type of radiation. Commercial communications satellites do not need radiation-hardened solar cells, if the solar devices are made of polycrystalline Si or GaAs or CdS or CdTe thin films. Military communications satellites must be protected from nuclear radiation effects using hardened solar devices or nuclear shielding. Regardless whether a communications satellite is commercial or military, damage to the satellite depends on the distance to and the magnitude of the nuclear explosion. If the satellite happens to be close to the nuclear explosion zone, one can expect catastrophic failure of the solar cells, leading to complete failure in the communication capability of the satellite.

6.7 Solar Array Design for Space Applications

The solar array configuration and design requirements must be given serious consideration to meet the specific performance and reliability specifications under harsh operating environments in space and over the stated operating life span. The operating environment in space includes low-energy proton and electron radiation belts, wider temperature excursions, and unpredictable illumination conditions.

6.7.1 Solar Array Design Requirements for Reliable Performance over a Specified Life Span

Selection of the type of solar array for space application is strictly dependent on:

1. Satellite orbit
2. Stabilization and attitude control
3. Electrical power output requirement over the operational life of the spacecraft
4. Communication satellite or surveillance/reconnaissance satellite type
5. Space available for the installation of solar arrays on the launch vehicle nose cone
6. Additional power requirements for stabilization and attitude control, which is dependent on the spacecraft stabilization scheme, such as a dual spin approach or three-axis stabilization technique

A schematic diagram identifying the orbit characteristics of spinning satellites is shown in Figure 6.10. This figure shows the approximate locations for the antenna, solar arrays, and equatorial orbit for a communications satellite. This is the most simple configuration for spinning communications satellites, which receive power from sun-oriented solar arrays. However, the sun-oriented solar arrays are only used when the electrical power requirements exceed the order of several hundred watts. In such cases, complex three-axis spacecraft stabilization is required and additional electrical power for the attitude control mechanism must be provided. It is important to point out that the use of sun-oriented solar panels provides nearly uniform power output throughout each orbit. The temperature of the sun-oriented solar panels when illuminated can vary from 60°C to 80°C, depending on the altitude and temperature at that altitude.

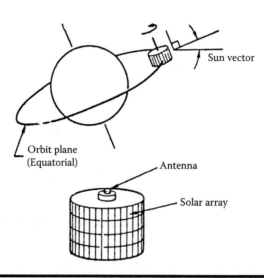

Figure 6.10 **Schematic representation of orbital characteristics and solar array location for a spinning satellite.**

6.7.2 Solar Array Orientation Requirements

The number of solar arrays and their orientation requirements depend on the orbit classification, total electrical power needed to provide electrical power to sensors and systems aboard the communications satellite or surveillance/reconnaissance spacecraft, and the power required for the attitude control mechanism functions. For equatorial orbits, a one degree of freedom array orientation will maintain normal solar energy impingement within a seasonal variation of ±23.5°. However, two degrees of freedom is necessary for other orbits where the spacecraft-sun line may be at any angle with respect to the solar array installation plane [1].

The orientation mechanisms or devices add additional complexity to the satellite or spacecraft system design specification. These orientation devices require a reliable way to transfer electrical power across the rotating joints, in addition to power needed for rotation. Since the electrical power requirements will grow with satellite mission functions, the tradeoffs will tend to favor oriented solar arrays and fully stabilized spacecraft necessary for excessive longevity and higher reliability. Preliminary studies performed by the author on solar array installation on communications satellites indicate that extendable lightweight solar arrays are best suited for space-based power systems. Note the size and weight of a solar array is strictly dependent on the electrical power rating of the solar array, number of modules in the array, electrical match between the series and parallel solar cells, density of the semiconductor film used in the fabrication of the solar device, and the types of solar modules deployed.

6.7.3 Electrical Power Output Capability of a Solar Array

The electrical power output capability of a solar array is dependent on the number of solar cells, conversion efficiencies of the solar devices, array orientation with respect to the solar energy impingement direction, solar mounting scheme, spin-stabilization technique, and the maximum allowable size of the array. Several hundred watts are possible from the body-mounted solar arrays generally used on spin-stabilized spacecraft. The choice between body-mounted arrays and fixed-paddle arrays depends on the mounting area available on the satellite or spacecraft body itself, the appendages of the satellite, and the overall weight requirements for maximum longevity and minimum launch costs [1]. Note minimum electrical power is needed for oriented solar arrays and for fully stabilized satellites or spacecraft.

It is important to mention that fixed paddle-type solar arrays can accommodate arrays with large effective areas for a given payload volume. However, such arrays will have a slightly lower power output capability per unit weight than the body-mounted arrays because additional structures are needed to provide structural integrity for the paddle-type arrays. Body-mounted arrays are best suited for most communications satellites and the solar devices are installed on the cylindrical surface as illustrated in Figure 6.10. For spinning communications satellites, the spin axis is generally normal to the orbit plane to assure optimum antenna gain. For spinning satellites in equatorial orbits, the body-mounted solar arrays with cylindrical configuration offer optimum antenna performance. Note this configuration offers about 8.5 percent lower solar power level than under the most favorable illumination conditions and only during the summer and winter when the sun angle is 23.5 degrees. Furthermore, the seasons of minimum solar array electrical power output for high-altitude orbits coincide with continuous daylight conditions. For inclined orbits the ends of the cylindrical must be covered with solar cells or the spacecraft shape must be modified during the periods when only one end of the cylinder is illuminated. This is very important particularly for satellites operating in polar orbits.

The electrical power output capability of a solar array is strictly dependent on the satellite payload requirement based on minimum launch cost and the maximum possible area available for installation of solar cells to meet specific electrical power consumption. For low earth orbit communications and TV broadcasting satellites, solar arrays with 5-kW and 12-kW electrical power ratings have been used widely since 1975. These solar arrays, illustrated in Figure 6.11, provide sufficient electrical power output over durations ranging from 5 to 7 years. Improved solar array designs using advanced materials and high-efficiency photovoltaic solar cells can provide longevity exceeding 10 to 12 years for space-based solar arrays. For space-based satellites to provide reliable target surveillance and reconnaissance missions may require solar array ratings in excess of 25 kW because of high power lasers, high resolution side looking radars and long range infrared tracking sensors deployed by the satellites. In the case of multichannel TV broadcast satellites and long-range covert communications satellites, rollout flexible substrate solar arrays are used to accommodate the large area with minimum launch cost.

6.7.4 Body-Mounted Solar Array Surface Temperatures

Excessive surface temperatures can degrade solar array electrical performance in terms of conversion efficiency and power output. Furthermore, the rise in surface temperatures is strictly dependent on whether the solar arrays are using body-mounted or fixed-paddle configurations and on the orbiting altitudes [1]. For body-mounted solar arrays, the surface temperatures of spinning communications satellites varies between 0°C and 20°C except when the sun vector is parallel to the spin axis, in which case the surface temperature of the continuously illuminated solar panels could reach about 80°C in earth orbit. Furthermore, the solar panel surface temperatures are slightly higher when the satellites are operating at lower altitudes because of the earth albedo or whiteness. Examination of atmospheric characteristics indicates that the atmospheric temperatures vary from about 270 K at 50 km to 210 K at 100 km operating altitudes. Note significantly higher atmospheric temperatures can be anticipated at higher operating altitudes exceeding 100 km. This means that solar panels installed on spacecraft or satellite will

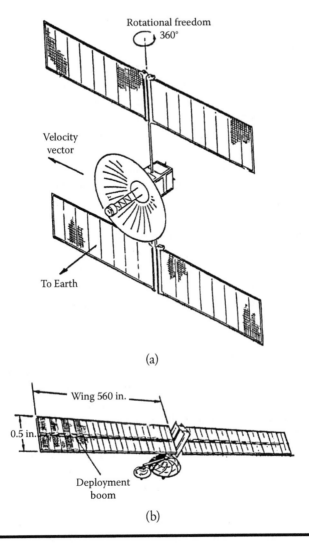

(a)

(b)

Figure 6.11   Solar array configuration for (a) a 12-kW direct TV broadcast satellite and (b) a 5-kW TV satellite.

encounter higher operating temperatures orbiting in medium or geosynchronous orbits. Under these circumstances, stringent material specifications must be defined for the construction of solar arrays to meet the electrical performance and longevity requirements, whether the solar arrays are employing the body-mounted or fixed-paddle configuration. It is important to mention that higher-channel/higher-date rate communications satellites or surveillance/reconnaissance spacecrafts are placed in synchronous equatorial orbits so that their apparent stations are continuously above the same points on the earth's surface all the time.

6.7.5 Mechanical Design Configurations for Space-Based Solar Arrays

Requirements for mechanical design and array configurations for space-based solar arrays are very stringent because of space radiation environments, harsh atmospheric conditions including temperatures and variable solar illumination densities. A solar array consists of a number of solar cells mounted on a substrate surface. Electrical interconnections and the electrical power-collecting buses associated with the solar devices complete the solar array structure. The solar cells are arranged in parallel- and series-connected strings to meet the desired current and voltage requirements. Blocking semiconductor diodes are deployed to separate the strings to improve the reliability of the array. The solar array design is contingent on the type of stabilization used by the satellite or spacecraft.

For a spin-stabilized communications satellite or surveillance spacecraft, the solar array is arranged in groups of parallel cells connected in series that are parallel-connected at their terminals. Note these groups or strings are arranged parallel to the cylindrical axis of the satellite. As far as the construction is concerned, the solar array can be fabricated as one unit or may be divided into a number of identical solar panels depending on the total panel surface area that is required to meet the electrical power output level. Regardless of the panel fabrication scheme, the solar cell groups must be bonded directly to the substrate, which typically comprises two face sheets of epoxy-fiberglass bonded onto a light-weight aluminum honeycomb core capable of providing high mechanical integrity under space environments. Appropriate adhesive materials must be used to mount the solar cell modules to the substrate, allowing for differential thermal expansion of the solar modules and the substrate. The differential thermal expansion aspect must be given serious consideration because the solar panels will be subjected to wide temperature variations under eclipse environments. It is important to mention that lighter solar panels and higher earth orbits will result in wider temperature excursions from sunlight to eclipse periods. Furthermore, wide temperature excursions along with the requirement for long operating life will put great demand on the design and reliability of the solar cell modules and intermodule connections in a solar array.

6.7.5.1 Design Requirements for Intercell and Intermodule Connections

Design requirements for the intercell and intermodule connections must be given serious consideration during the solar panel layout and fabrication of the array. The material and form of the interconnections between the individual solar cell modules must be carefully selected with particular emphasis on the temperature cycling because of its impact on solar array reliability and longevity in unpredictable space environments. Material science studies performed by the author indicate that molybdenum and Kovar materials are best suited for interconnections. These materials have thermal expansion coefficients that will match silicon quite well under wide temperature excursions. In case of spinning arrays or low earth orbit satellites, silver or copper can be used for interconnections because of less temperature extremes. However, future communications satellites will use welding techniques to eliminate the solder weight and to improve mechanical integrity and solar array reliability over extended space operations. In summary, a solar array fabrication typically involves positive and negative metallization, low-cost glass cover, intercell connections, inter-module connections, Kapton dielectric, and silicon base material.

All solar cells must be protected with cover slides. The solar module is the smallest subassembly consisting of solar cells and interconnects. It is self-supporting during the handling sequences of the array assembly and can be fully inspected before it is bonded to its substrate. This visual inspection also meets quality control and assurance specifications, which are of paramount importance for systems operating in space environments. The cover slide made of fused silica with thickness ranging from 0.003 to 0.020 in. offers protection from proton and electron radiations encountered in earth's orbits. The cover slide must meet the radiation resistance, cost and weight requirements, and must be compatible with the cost-effective design, reliability, and longevity specifications. Materials such as sapphire and micro-sheet glass are available for the cover slide, but because of its low cost and mechanical and thermal stability in space, fused silica is preferred. The protection cover slide has antireflection coating on one side to minimize reflection losses at wavelengths from 500 to 800 nm and ultraviolet reflectance interference filters on the other side to provide protection of the cell-to-slide adhesive under ultraviolet (UV) radiation. Radiation damage to solar cells can be caused by charged-particle environments, which include trapped radiation and solar flares. Regardless of the radiation source, the radiation can affect the transmittance of the solar cell cover slides and cell-to-cover adhesive and the diffusion length of the minority carriers in the semiconductor materials, resulting in performance degradation of the solar cells. The intensity and the spectra of proton and electron radiations vary widely with orbit altitude, satellite inclination, and solar activity. Essentially, the cover slide is the major means to protect the solar cell from the adverse effects of space environments.

Adhesive made from epoxies and silicones are widely used for space-based solar arrays, because of their superior ability to withstand ionizing radiation for very long durations without any structural changes or any degradation in optical transmittance. Note any increase in solar absorption will result in increase in the solar array operating temperature, which will reduce the electrical output of the array. It is important to mention that the UV degradation of optical properties of the adhesives must be minimized to achieve the minimum power output of the solar array over the entire operating life of the solar power system. Data from previous communications satellites indicate that significant coloration of adhesive due to charged-particle irradiation must be prevented to ensure the reliability and longevity of the space-based solar arrays.

6.7.5.2 Sources of Weight Contributions to Solar Arrays

The weight of a solar array is the critical design parameter, which partially determines the launch cost and the operating life of the satellite or spacecraft. Several factors contribute to the weight of the solar array, such as size of the array, interconnections, cell-to-cover adhesive, soldering material, protection cover, solar cells, support accessories, solar panel installation parts, and structural materials. Reduction of solar array weight must be given serious consideration, if launch cost and longevity are the principal design requirements. Studies performed by the author on various solar array configurations indicate that arrays using the rollout flexible substrate technology will offer minimum weight. The studies further indicate that a honeycomb structure will not only yield lower weight, but also higher mechanical strength under wide temperature excursions in space. Studies performed by the author on solar cells using thin-film technology reveal that thin-film cell construction offers high conversion efficiencies, which in turn yields lower array weight and higher electrical power to mass ratio, irrespective of solar illumination intensity.

6.8 Summary

This chapter has focused specifically on the design requirements for solar cells and arrays for space applications. Material requirements for space-based solar cells and arrays are summarized. Important electrical, thermal, and mechanical properties of Si and GaAs materials are discussed for space-based solar devices. Performance parameters of CIS, CIGS, InP, and CdTe are highlighted, with emphasis on conversion efficiency and space radiation resistance. Performance capabilities and limitations of Si and GaAs solar cells are identified. Performance, cost, and weight aspects for selected solar cells are discussed, with particular emphasis on space radiation resistance and longevity. Adverse effects of high-intensity proton and electron radiation levels on solar cell performance are summarized, with major emphasis on catastrophic failure of the solar cells. Reduction in open-circuit voltage and conversion efficiency of solar cells operating in communications satellites due to space radiation in low earth's radiation belts is discussed in great detail. Low-energy proton damage in ion-implanted and diffused solar cells is identified. Adverse effects of wide temperature excursions in space on open-circuit voltage and conversion efficiency are summarized. Reduction in open-circuit voltage in heterojunction GaAs solar devices as a function of temperature excursions and operating hours is discussed, with emphasis on device performance level. Performance capabilities and limitations of high-efficiency multijunction and multilayer solar cells are summarized. Performance characteristics of dual-junction and triple-junction solar cells using heterojunction structures are identified, with major emphasis on conversion efficiencies, which range between 21 and 25 percent at room temperature and under laboratory conditions. Performance characteristics of triple-layer

GaInP/GaAs/Ge solar cells are summarized along with particular advantages due to enhanced absorbing capabilities over specific solar spectral regions. Trends in solar cell conversion efficiencies at the beginning of life (BOL) and end of life (EOL) are clearly identified. Design configuration aspects for space-based solar cells and solar arrays are discussed briefly, with emphasis on longevity. Electrical power output capabilities for space-based solar arrays as a function of film thickness and array area are summarized. Critical mechanical design aspects of solar arrays are discussed, with emphasis on survivability with no catastrophic failures of solar cells. Solar array design requirements for spin-stabilized communications satellites are identified with major emphasis on weight and cost. Cost-effective techniques for intercell and intermodule interconnections are proposed. Weight reduction techniques for solar arrays are proposed because heavier arrays will increase installation complexity and satellite launch costs.

References

1. W.E. Berks and Werner Luft. "Photovoltaic solar arrays for communications satellites," *Proceedings of IEEE* 59, No. 2 (1971): 263–373.
2. J.F. Loferski. "Recent research on photovoltaic solar energy conversion," *Proceedings of IEEE* (1963): 667–673.
3. Alan G. Stanley. "Comparison of low-energy proton damage in ion-implanted and diffused silicon solar cells," *Proceedings of IEEE* (1971): 721.
4. R.C. Knechtli, G.S. Kamath et al. "High-efficiency GaAs solar cells," *IEEE Transactions on Electron Devices* ED-31, No. 5 (1984): 577–587.
5. A. Goetzberger and V.U. Hoffmann. *Photovoltaic Solar Energy Generation.* New York: Springer, 2007.
6. Nasser H. Karam, Richard King et al. "Development and characterization of high-efficiency GaInP/GaAs/Ge dual- and triple-junction solar cells," *IEEE Transactions on Electron Devices* EC-46, No. 10 (1999): 2116–2117.

Chapter 7

Design Requirements for Stand-Alone and Grid-Connected PV Systems

7.1 Introduction

Design requirements for stand-alone photovoltaic (PV) power systems and grid-connected photovoltaic (PV) power systems are discussed in this chapter. The power output capability of a solar power system is dependent on several considerations regardless of which PV system is used. For example, the minimum electrical power level needed, the geographical location, and the average sunlight available per year are the three major considerations. For a stand-alone PV power system a solar collector or solar panel and a standby battery are needed to provide electrical power 24 hours continuously. The combination of least expensive solar modules and storage batteries offers a continuous, year-round operation for a certain electrical power requirement. But in the case of a grid-connected power system, the solar arrays and a grid connection from the commercial electrical utility company are required to meet the consumer electrical power requirements on continuous basis. A solar heating system represents a classic example of a stand-alone PV power system, which provides electrical energy for hot water and for heating residential homes and business buildings. A residential owner who is only interested in acquiring the electrical energy from the solar arrays demonstrates a good example of the grid-connected PV system because the house is connected to an electric utility grid. It is important to mention that there are several applications of stand-alone PV power systems.

Computer programs are available to select the smallest solar array size for a given electrical load requirement based on continuous availability of solar energy, the storage battery operating parameters such as battery discharge rate, charge efficiency, maximum battery discharge allowed and annual battery self-discharge, and optimum tilt angle for the solar panels with most uniform output over the entire year. Based on the electrical power consumption requirements, the computer program will determine a least-expensive solar array design, number of solar modules required, and the storage battery specifications. In 1992, aerospace scientists presented the concept of a "solar power satellite" (SPS) system [1], which will convert the solar energy in space for use on earth. This SPS system converts the solar energy directly into electricity and feeds it to microwave generators forming part of a planar, phased-array transmitting antenna. This high-gain, electronic steering antenna would direct a microwave beam to one or more high-gain receiving antennas at appropriate locations on earth. This high power microwave energy would be efficiently converted into electricity, which can be transmitted and distributed for customer use. This SPS system will be discussed later on with emphasis on reliability and economics.

7.2 Grid-Connected PV Power Systems

A grid-connected PV power system integrates the electrical energy obtained from PV solar cells located either on a tower, residential roof, or commercial building into a public utility electric company. Grid-connected PV power systems can be classified as follows:

- The owner of a house, commercial building, or factory who has installed solar arrays on the roof, where solar arrays are generally installed, represents a classic example of a grid-connected PV power system. The PV cells are exposed to solar incident energy, which converts the solar energy into electrical energy [2] needed to operate lights and electrical and electronic appliances. In a grid-connected PV power system, storage batteries are not needed to provide continuous, year-around operation, because the grid provides the electric power in the absence of sunshine.
- Discount stores in California, such as Target and WalMart, have installed solar panels to meet their electric power requirements in conjunction with the electrical power supplied by the grid of the Southern California Edison.

When sunlight is available, the stores use electrical energy derived exclusively from the sun.

7.2.1 General Description of a Grid-Connected PV System

A grid-connected PV electrical power system, when properly designed and installed, will be capable of meeting the consumption power requirements of a television set, refrigerator, toaster, hair dryer, radio, washer and dryer, window air-conditioning unit, electrical lights, and other electrical appliances. Central air-conditioning with a 3-ton rating will require a PV system capable of generating additional electrical power close to 11 kW to meet the requirements of all these appliances. A grid-connected solar power system consists of light-weight solar panels, inverters with long life and high efficiency, heavy-duty mounting rails, mechanical support accessories, and electrical cables and wiring leading from the solar panel output to the electrical meter installed by the company. It is important to mention that solar panels or modules, mounting rails, and inverters are the high-ticket items. The remaining accessories and components for the grid-connected solar system would cost less than 10 percent of the total system cost.

7.2.2 Roof-Mounted Solar Panel Installation Scheme and System Cost Breakdown

Roof-mounted solar panel installation has been preferred because of the low cost of installation with minimum effort. The panels can be made with optimum length and width, minimum weight, and high reliability. Typical roof-mounted solar panels are illustrated in Figure 7.1 (a). The panels do not have to be installed flat on the roof surface and can be installed at an angle of inclination best suited to achieve optimum electrical performance. The panels must be installed facing east to get solar energy over longer hours of the day. The panels can be mounted on a light platform with two-dimensional sun tracking capability to convert maximum solar energy into electricity. However, the cost for solar panels mounted on a sun tracking platform will be higher [3].

Solar panels currently available on the market have typical dimensions of 5 ft by 3 ft, with typical electrical ratings or output levels ranging from 180 to 215 W. Higher ratings are possible with larger panel dimensions, but at the expense of additional weight. A typical 200-W solar panel costs between $1000 and $1100. Panels with lower electrical ratings cost slightly less. Typical panel life varies from 25 to 30 years, according to commercial panel suppliers. Inverter cost is less than panel cost and comes with a life warranty of 10 years. Glass covers are used to protect the panels from dust, rain, snow, and sandstorms. Based on preliminary estimates, a grid-based solar system with a 5-kW rating would meet all electrical consumption requirements for a family of four, except for electricity needed to operate a central air-conditioning with moderate rating [3].

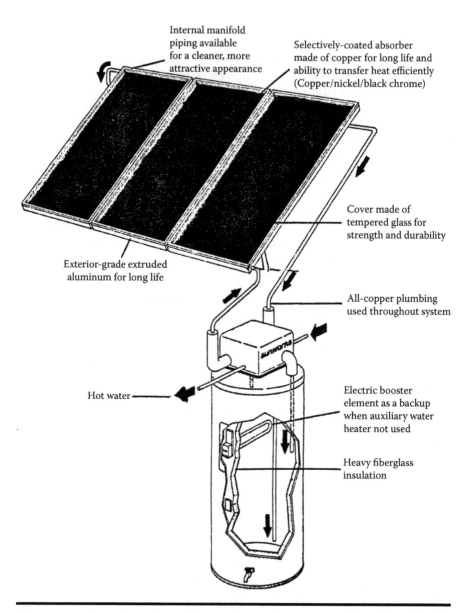

Figure 7.1 Layout of a commercial solar hot water system for domestic use.

7.3 Stand-Alone PV Power Systems

A wide range of potential applications for stand-alone PV power systems include:

- Communications satellites
- Surveillance and reconnaissance spacecraft
- Navigation satellites
- Radio relay stations
- Navigation aids in remote locations
- Unprotected railway crossings
- Emergency location and alarm transmitters
- Educational TV programs broadcasting in remote areas
- Airport landing obstruction lights
- Cathodic protection systems for corrosion control
- Water pumping for drinking and irrigation
- Power for fire observation towers and second homes
- Electric border fences and intrusion warning alarms
- Marine battery and other float charge systems
- Highway signs for motorists
- Electronic watches and calculators
- Portable backpack radios and covert communications systems

7.3.1 Design Configuration and Critical Performance Requirements for Stand-Alone PV Power Systems

Performance requirements and design parameters are strictly dependent on the type of stand-alone power system. For example, design and performance requirements for a water heater will be different than those for a swimming pool or those needed exclusively for heating a building such a school or gymnasium or office. For a domestic water heater or a swimming pool, one requires a quite simple PV power system. For heating a large building, a costly and complex PV power system with exotic control and monitoring accessories will be required. Computer-aided design (CAD) can be used to determine the best PV power system available in terms of electrical load and geographical location for a specific stand-alone PV power system.

7.3.1.1 Water Heater Design Using Solar Technology

Solar technology offers the most economical and environmentally friendly approach for water heating for domestic consumption. On a typical sunny day in southern California or in similar regions around the world, solar radiation provides energy of more than 200 BTU per hour per square foot, representing a significant source of energy at no cost. Currently, high-technology solar collector design offers operating efficiencies exceeding 70 percent [4]. This translates into a solar energy of 6400 BTU per hour available from a 4 ft × 8 ft solar collector. Assuming a solar collector efficiency of 70 percent, an absorbed energy of 4480 BTU/hour is available from a 4 ft × 8 ft solar panel, which amounts to 1.31 kWh of electrical power available, because 3413 BTU is equal to 1 kWh. To recover this amount of electrical energy in a typical installation requires a 100-W circulation pump.

It is important to point out that water heating for household use requires an enclosed unit to insulate it against heat loss due to large temperature differences between the ambient air and the solar panels or collectors, which could be as much as 100°F particularly in the winter season. Physical layout of a commercial water heater [4] for domestic use is shown in Figure 7.1. The 4 ft by 8 ft solar collector appears to be most satisfactory for small households and is widely available in the market for home use. However, for large households two or three solar panels may be required, as illustrated in Figure 7.1. Solar panels or collectors that are 4 ft by 10 ft are commercially available, which offer better economical operation. However, they are proportionately larger and heavier compared to 4 ft by 8 ft solar collectors and could increase the installation costs. The most appropriate and ideal location for the solar collector is the roof of the house or building.

7.3.1.2 Description of Critical Components of the Solar Hot Water System

The most important and costly component of a solar hot water system is the solar panel or the solar collector, which is illustrated in Figure 7.2. Specific structural details of a commercially available solar collector are shown. The aluminum frame provides the mechanical support for the serpentine-connected copper tubes carrying

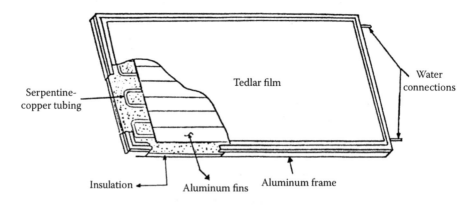

Figure 7.2 Structural details of a solar panel best suited for use in a hot water system.

the water to be heated from the sunlight. Aluminum fins are provided for uniform heating. Inlet and outlet water connections are shown along the smaller dimension of the panel to avoid heat loss. As stated earlier, for a larger household unit two or more solar collectors may be required, as illustrated in Figure 7.1.

The solar collector must be installed at an appropriate location on the roof, which must provide maximum exposure to the sunlight for extended hours of the day [4]. Internal manifold piping must be provided for a cleaner and attractive appearance of the collector. An efficient heat absorber made from copper/nickel/ black chrome composite media must be used for the most efficient heat transfer over the entire life of the collector. Exterior-grade extruded aluminum at the edges as illustrated in Figure 7.1 is provided for the longevity and higher mechanical integrity of the solar collector. A tempered glass cover is provided for safe operation, structural strength, and durability. All-copper plumbing is recommended throughout the heater for longevity, maintenance-free operation, and clean water supply. Heavy fiberglass insulation is absolutely necessary, if prevention of heat losses over the life of the system is desired. The directions of cold and hot water to and from the water heater are clearly shown in Figure 7.1.

Technicians and mechanics who install solar panels must have previous experience and demonstrate reliability and good workmanship. Installation of solar panels with sloppy tolerances, thin light duty mounting rails, and inverters with excessive weight would require many more roof penetrations, which could lead to damage to the roof over time. There could be water leakage, thereby affecting the mechanical integrity of the roof.

7.3.1.3 Cost of Domestic Solar Water Heaters

The cost of a domestic solar water heater is a function of hot water consumption per day for daily domestic services such as washing clothes, showers, tub baths, dish washing, and other services. The acquisition cost for a commercial water heater is still quite high. Current estimates vary from $3000 to $5000 depending on consumption requirements. It could take more than 5 years to make the break-even point considering the savings in operating costs. There have been claims that solar water heaters have lifetimes greater than 20 years provided recommended maintenance procedures are strictly followed. There is a possibility of some damage to the roof due to water leakage from the pipes in the solar panels, if the solar modules or panels are not properly installed. Home owners are advised to ask the installing companies to be liable for any damage to the roof from water leakage. Customers must contact potential solar panel installers on specific performance parameters and warranties on solar panels and inverters, prior to signing any contract for solar system installation. For higher longevity of the solar panels, inverters with operating life greater than 10 years must be selected.

7.3.1.4 Federal and State Tax Incentives for Solar System Installations

There are federal and state tax incentives for installation of solar panels, whether they are intended for converting solar energy into electrical energy or for hot water application. The federal tax incentives are uniform within the continental United States, Hawaii, and Alaska. State tax incentives vary from state to state. According to Solar Home Product Inc., the state of California just dropped the commercial solar rebate from $2500 in 2007 to $1800 per kilowatt in 2008. However, the solar rebate for residential customers currently in California is $2500 per kilowatt.

For a limited time, a 30 percent federal tax credit was in effect for residential customers, but there was no cap for commercial installations. However, this federal tax credit expired at the end of 2009, according to Solar Home Product Inc.

7.3.1.5 Estimation of Solar Collector Area Needed to Meet Hot Water Consumption Requirements

Design of a solar water heater is dependent on the hot water consumption, which is a function of various services requiring hot water such as bathing, washing, and so on. Hot water consumption in summer will be significantly less than that required in summer. Typical hot water consumption for a family comprising four persons would require a minimum of 40 gallons per day for the above activities [3]. It is extremely difficult, if not impossible, to estimate accurately. Similarly, the savings that one can realize from solar heating for pool and /or water heating are equally difficult to predict with great accuracy. Commercial solar water heater suppliers could provide the best estimates for both the hot water consumption requirements for a given family size and solar pool heater. During the swimming season, solar heater design is strictly a function of the ratio of the solar collector area to the pool area. Swimming pool heating suppliers indicate that one can save about $200 to $250 per year on gas heating bills, provided the pools are exposed to moderate temperatures during May to mid-November, with water temperatures not exceeding 80°F in mid-summer. In general, solar pool heaters are not running all the time. Savings from solar pool heaters are dependent on family size, climatic seasons, and comfortable temperature settings. In southern California, one can greatly reduce gas consumption for 8 to 10 months.

7.3.1.6 Design Requirements and Description of Solar Collectors

The solar collector is the most critical element of a hot water system, regardless of applications such as providing hot water for a home or heating a swimming pool. Critical elements of a solar collector for a hot water system and for a swimming pool are shown in Figure 7.2 and Figure 7.3, respectively. In each case, structural details and critical components such as the header, collector fins, copper pipe sections,

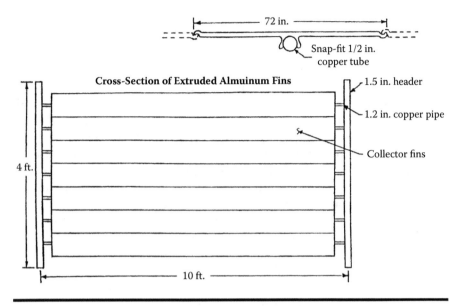

Figure 7.3   Plan of a solar panel used widely for swimming pools.

and extruded aluminum fins are clearly illustrated. The exterior-grade extruded aluminum section as displayed in Figure 7.2 is used for longevity of the solar collector. All-copper plumbing throughout the hot water system is used to provide clean water and to prevent corrosion in the pipes over the life of the system.

The basic solar collector or panel design consists of copper waterways needed to extract solar heat from the extruded aluminum collector fins, as shown in Figure 7.3. These fins are snap-fitted over the copper pipe waterways. For the swimming pool solar heater, the copper pipes are connected to a common manifold at each end. For the solar water heater, a serpentine connection is employed, as illustrated in Figure 7.2. The collector assembly or the solar panel assembly is coated with a flat, highly absorptive coating to absorb maximum solar energy from the sun. The collector assembly is semi-open for swimming pool heating, but must be enclosed for domestic solar water heating. The enclosure generally consists of an insulation backing of high temperature closed cell foam, aluminum frame, and a top cover of high strength Dupont Tedlar film as shown in Figure 7.2. Note the Tedlar film is not susceptible to breakage as glass is. The Tedlar film has better light transmission properties than 1/8-inch window glass and can be easily patched if accidentally punctured. According to Dupont scientists Tedlar film will maintain at least 50 percent of its original tensile strength after 15 years. To maintain higher tensile strength after 20 years, one must select appropriate film material with excellent light transmission properties.

Solar collectors or panels come in three sizes:

■ 4 ft. × 8 ft. glazed (net area available = 31 sq. ft.)

- 4 ft. × 10 ft. glazed (net area available = 39 sq. ft.)
- 4 ft. × 10 ft. unglazed (net area available = 39 sq. ft.)

The 4 ft. × 8 ft. collector or panel size is commercially available at various locations and is best suited for house use. It is easy to install on a wide variety of roof styles and shapes. Panels with dimensions 4 ft. × 10 ft. are also available, which cost somewhat less per square foot than the 4 ft. × 8 ft. panel. However, they are proportionately larger and heavier and could increase the installation costs.

Correct estimation of the collector area is dependent on several factors, such as hot water requirement for bathing in terms of gallons per person per day, climatic seasons (winter or summer), hot water for washing clothes, hot water for dishes, and number of persons in the family. Hot water consumption is generally lower in summer than in winter. Assuming an average hot water consumption of 15 gallons per person per day and one square foot of collector area per gallon, the collector area requirement for a family of four comes to 60 sq. ft. This means the storage capacity for a hot water system for a family of four will be 60 gallons.

Critical elements of this type of hot water system configuration are shown in Figure 7.4. The layout shown in Figure 7.4 uses the existing water heater for storage.

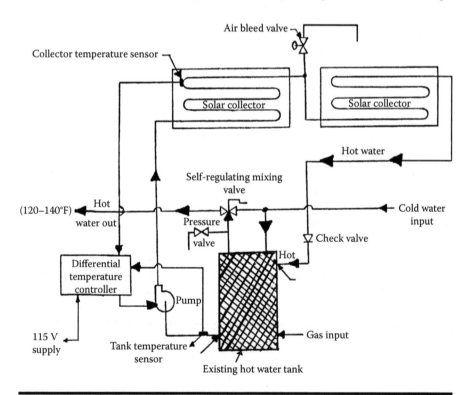

Figure 7.4   Details of a solar installation for a domestic water heater.

This system configuration requires two solar collectors or panels, each of which occupies an area of 4 ft. × 8 ft which is equal to 32 sq. ft. In addition to the two collector panels and water heater, the installation requires a small-capacity circulating water pump and an automatic controller. The controller is the most critical component of the installation. The controller only permits the circulating pump to circulate water through the solar panels exposed to sun, when the panel temperature is higher than the water in the storage tank. This prevents the heat loss due to pumping hot water through a cool panel, thereby giving maximum water temperature. One can expect the water temperature in the tank to reach between 160°F and 180°F during the summer or warmer months. The self-regulating mixing valve shown in Figure 7.4 is used to mix the cold water in before use, so that the water delivered to the faucets will be in the normal temperature range of 120°F to 140°F, which is considered sufficient for taking baths or washing clothes. This technique increases the overall efficiency of the hot water system by minimizing the heat loss in the plumbing of the house [4]. The heat loss is a function of the temperature difference between the air and the pipes. The higher the temperature difference, the more rapid will be the heat loss. The overall efficiency can be further improved by insulating hot water pipes wherever possible. It is essential to insulate the supply and return lines to the solar panels installed on the house roof as illustrated in Figure 7.4.

An additional or auxiliary storage tank is strongly recommended, if the existing tank does not meet the hot water requirements or it has a marginal storage capacity. A typical hot water installation system comprising two storage tanks is shown in Figure 7.5. This particular installation scheme has several advantages over replacing the current hot water system with a larger unit. According to hot water tank suppliers such as Sears, Home Depot, etc., addition of a second tank will cost less. By adding an 82-gallon tank to the current 40-gallon tank, one can increase the hot water storage capacity to 120 gallons for about $150 or so. The additional storage capacity will help to ensure hot water is available on cloudy days or to meet the additional hot water requirements for laundry or other domestic services.

Another advantage over a single large heater comes in when there is insufficient solar energy to meet all hot water requirements. Under these circumstances, the conventional water heater, which is set at the lowest temperature, comes on and makes up the difference in temperature. With one large water heater, one has to heat the entire tank of water, but with the two tank scheme as illustrated in Figure 7.5, heat is required only for the 40 gallon tank associated with the existing water heater installation, thereby reducing the amount of gas needed to heat the water. This hot water installation offers savings of about $800 in gas heating over a period of five years for a family of four.

7.3.1.7 Cost Estimates for a Typical Hot Water System

The hot water system configuration illustrated in Figure 7.4 is widely used for a family comprising four or fewer family members. This system configuration uses

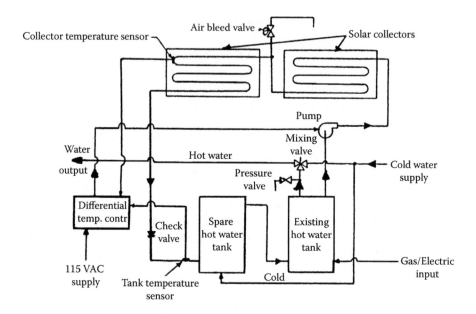

Figure 7.5   Alternate solar installation for domestic hot water.

two solar collectors, one hot water tank, one collector temperature sensor, one mixing valve, one pressure valve, one circulating pump, one hot water tank temperature sensor, one check valve, and one differential temperature controller. Based on the information provided by the suppliers, estimated procurement costs for the above items can be summarized as follows:

- Two 4 ft. × 8 ft. glazed solar collectors at $200 each = $400
- One stainless steel circulating pump = $150
- One check valve = $15
- One mixing valve = $25
- One differential temperature controller = $135
- 1/2-inch copper pipe and fittings = $100
- Miscellaneous accessories, such as hoop-wire, insulation material for pipes, and small hardware needed for installation and mechanical support = $100
- Estimated total parts cost = $925, plus tax

The overall hot water system cost must include the estimated part cost plus tax plus installation cost. The installation costs vary from city to city and will be higher for a licensed installation mechanic compared to a handyman installer. Once all the necessary parts, city permit, and the installation mechanic are available, the hot water system installation can be completed within two to three days, if the city inspectors complete the inspection process without delay. As far as the payback period or breakeven time for a domestic hot water system is concerned, one can recover all the installation and part costs within five years after the installation date, assuming the cost of gas does not rise more than 8 percent per year over the five year duration.

7.3.2 Closed-Loop Active Hot Water System Using Solar Technology

The closed-loop active water heater using solar technology is best suited for locations subject to frequent or prolonged freezing weather or where the water supply is hard and results in severe scaling. Typically solar panels or collectors in a conventional system are installed on the roof, but in the case of an active hot water system an option is available to locate them elsewhere because the liquid is distributed by the pump as illustrated in Figure 7.6. The hot water system depicted in Figure 7.6 is classified as "closed loop" because the antifreeze liquid is pumped to the solar collectors down to a heat exchanger and back to the collectors in a closed circuit. The system is completely sealed off from the household water supply. The antifreeze liquid is used the eliminate the possibility of freezing within the collector pipes and to keep the hard water from scaling and reducing the absorbing efficiency of the solar panels without degrading the heat transfer efficiency of the system. The antifreeze is pumped up to and through the solar collectors and from the collectors to a heat exchanger. The heat exchanger makes the use of two different liquids possible.

Both the antifreeze and the household water flow through the heat exchanger but in separate pipes or tubes. Through a process of conduction, the solar-heated antifreeze heats the walls of the copper pipes, which also heat the water through conduction on

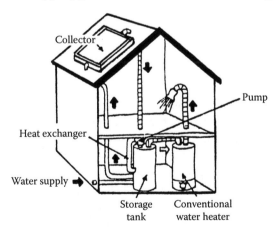

Figure 7.6 **Closed-loop solar system with roof-mounted solar collector, which circulates an antifreeze liquid and transfers heat to the household through conduction.**

the other side. The heated water is then stored in the preheat water tank. Both the antifreeze and the water are pumped only when there is a temperature difference of four degrees between the antifreeze in the collector pipes and the coolest water in the storage tank bottom. This temperature differential means it is time to circulate the solar-heated antifreeze liquid. Temperature sensors, one in the solar panel and one on the bottom of the storage tank, alert the differential temperature controller to the temperature difference, which sends a turn-on signal to the pumps. The temperature controller keeps the pumps turned off at night or during cloudy or freezing weather. An absorber plate comprising two sheets of pure copper is used to increase the heat transfer efficiency of the system. The channels carrying the liquid are closely spaced over the entire surface of the absorber so that the solar energy striking the solar collector surface has only a short distance needed for optimum thermal performance.

7.3.2.1 Major Component Requirements for a Closed-Loop Hot Water System

There are 11 major components [4] that control the liquid flow or distribution within the closed-loop active hot water system; they can be described as follows:

- High efficiency solar collectors (two)
- Preheat storage tank (one)
- Circulating pumps (two)
- Hot and cold sensors (as needed)
- Differential temperature controller with digital readout capability (one)
- Mechanical shutoff valves (as needed)
- Electrical solenoid valve (one)
- Flow indicator (one)
- Expansion tank (one)
- Heat exchanger (one)
- Copper piping (as needed)

The differential temperature controller with digital readout capability is the most critical component in closed-loop hot water systems. The controller protects against freezing and monitors the temperature differential for the domestic hot water system. This component is specially engineered for hot water systems requiring one differential temperature control output signal and a separate output providing the drain-down freeze protection. A three-position mode switch is a part of the controller module, which allows fast system performance check to provide fool-proof and safe operation. Independent thermostat sensors may be wired in series with the "collector" or "freeze" thermistor to provide multipoint freeze detection capability. LED indicators for red and green lights for assigned operating functions are provided in the controller assembly. Drain valves and relays are integral parts of the controller assembly. The "snap switch" sensor, which provides secondary freeze protection capability for the controller module, can be mounted in an appropriate outdoor location using piping insulation or attached to the absorber plate located on the solar panel.

7.4 Solar Heaters for Swimming Pools

Solar heaters for heating swimming pool water offer a cost-effective and very inexpensive alternative to costly electricity from a public utility company. Heating water for household use requires an enclosed heater system to insulate it against the heat loss due to larger differences in temperature between the ambient air and the solar collectors or panels. This temperature difference could be as much as 100°F in winter. Solar panels for heating swimming pool water can be installed on a floating surface in the water. In other words, the solar energy converted into electricity can be used instantly to heat the pool water without heat loss.

7.4.1 Solar Panel Requirements for Pool Heating System

Solar panel fabrication and size requirements for swimming pool systems are quite different from those needed for other solar applications. Solar panels 4 ft wide and 8 ft long are widely used for pool heating systems. Solar panels 4 ft by 10 ft are also available for this application, which cost somewhat less per square foot than the 4 ft by 8 ft panel. The larger panels are slightly heavier as well, but they can be installed in a relatively in short time. Lighter materials must be used in fabrication of solar panels for the swimming pool. A circulating pump with appropriate capacity, water filters, and temperature monitoring devices are required to maintain filtered water with uniform temperature within the pool.

Because considerably lower temperature is involved for pool heating, the solar panels for such applications do not have to be glazed to prevent excessive heat loss. Unglazed solar panels are quite capable of maintaining the swimming pool at 90°F during the summer. According to solar pool heater designers, water temperature between 75°F and 80°F over most of the swimming season is considered tolerable, but water temperature must be maintained around 90°F during the winter season.

It is recommended that the solar panel area be at least 50 percent of the pool area for a pool located in a moderate climate. For a 450 sq ft pool area, one must use solar panels with area equal to 225 sq ft. This means that as a minimum six 4 ft by 10 ft solar panels, as shown in Figure 7.7, are required for this pool size for 70 percent of the swimming season. However, nine such solar panels will be required for year-round swimming. Critical elements of a typical solar pool heater design configuration are depicted in Figure 7.6. Six unglazed solar panels, one check valve, one control valve, pool filter, heater module, electrical accessories, copper pipes, and control panel are required for such an installation. PVC pipes should be given serious consideration, because they will reduce both the panel weight and corrosion inside the pipes. Structural details of a solar panel best suited for pool heating have been

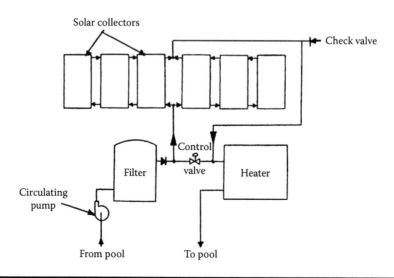

Figure 7.7   Solar swimming pool heater installation.

clearly illustrated in Figure 7.3. The structure includes fasteners, gaskets, back insulation layer, aluminum foil, side insulation layer, and aluminum retaining frame to provide the enhanced mechanical and thermal performance levels [4]. The solar collectors are extruded aluminum fins to improve overall panel performance. The solar panels must be well secured on the roof without damaging the structural integrity of the roof due to water leakage from the copper or PVC pipes used by the panels.

7.4.2 Operational Requirements of a Solar Swimming Pool Heater

Operational requirements of a solar pool heater are not very stringent compared to those for domestic solar water heaters. Few accessories and control devices are required to install the system, as illustrated in Figure 7.7, and no supervision is needed for the operation of the system. This particular system uses a minimum number of components, including a check valve, control valve, water filter, circulating water pump, and a heater, besides six solar collectors, as shown in Figure 7.7. If the water filters are not performing efficiently, they must be changed, which do not require any expert maintenance technician. The control panel will display electrical supply parameters and pool water temperature.

7.5 Tower Top Focus Solar Energy Collector System

The tower top focus solar energy (TTFSE) collector system is the most efficient and cost-effective stand-alone photovoltaic solar system. The critical elements

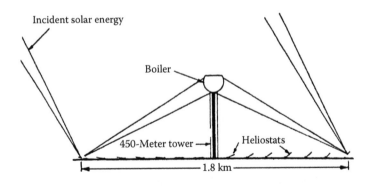

Figure 7.8 Cross-section of a tower top focus solar energy collector system capable of generating very high thermal energy.

of the TTFSE collector system are shown in Figure 7.8. This system consists of focusing mirrors, a black-body solar collector, boiler and steam turbine, which can be mechanically coupled to an electric generator capable of generating electrical energy in excess of 250 MW per day [5]. This TTFSE collector system has the highest capacity to collect the solar energy. This system offers a sophisticated technology capable of reflecting the sun's energy with mirrors to an elevated central point where it is efficiently absorbed and converted to heat by a black-body solar collector. The system offers high concentration ratios, leading to an overall conversion efficiency exceeding 30 percent and temperatures ranging from 550°C to 650°C. This high temperature in the boiler permits generation of superheated steam to drive a high-speed steam turbo-alternator capable of producing AC electrical power exceeding 250 MW. This thermal energy power system offers electricity generation cost per kilowatt-hour (one kilowatt-hour = 3413 BTU) that approaches a competitive price with oil and gas.

7.5.1 Operating Principle of the TTFSE Collector System

A cross-sectional view of the TTFSE collector illustrates the location of the boiler, tower details and an optical array consisting of low-cost flat mirrors capable of focusing the optically concentrated solar energy to the boiler located at the top of the tower. The heliostats [5] as shown in Figure 7.8 are the critical components of the system and provide very high concentration ratios exceeding 100:1. The solar energy irradiating 1 km² of ground at noon on a clear day south of 35° N latitude generates over 400 MW of heat even in midwinter. The larger the ground span the higher will be the amount of heat generation. This amount of heat is possible using optically concentrated solar energy and a thermodynamic cycle principle. The boiler height is about 450 m for a ground distance or span of 1800 m, which in this case provides an inclination angle of 27 degrees between the ground edge and the boiler location.

This optical array consisting of several mirrors is automatically steered to redirect the incident solar energy to a central focal point. It is necessary to space the mirrors carefully to minimize shading effect of one mirror by another and to have the collection point elevated high above the mirrors. As stated earlier the concentrated solar energy is absorbed and converted to heat energy by a black-body solar collector placed in the focal region. The heat can be transported down the supporting tower via liquid metal and/or steam lines and can be stored or used to operate a conventional turbine generator station.

7.5.2 Heliostat System Configuration

A typical heliostat system configuration is shown in Figure 7.9 and its essential components include sensor housing, mirror, elevation and azimuth control circuits, counter weight for static balance, and mechanical support. A sighting alignment hole plays a key operational role. A two-axis heliostat control is necessary for smooth and efficient system operation [5]. The heliostat control can be obtained by using either a hydraulically or electrically operated servomechanism, which drives a signal from a simple position-sensing phototube device. The heliostats typically range from 3 to 10 m² are spaced over a field with radius equal to one to two times the tower height in such a manner that they are largely visible from the receiver to minimize shading as the sun crosses the sky. Solar collectors bearing an obvious

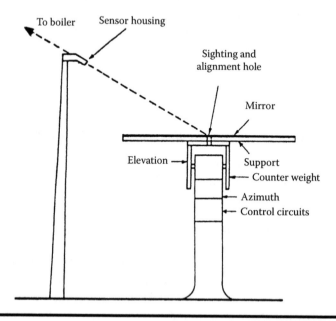

Figure 7.9 **Schematic diagram of a typical heliostat system showing the critical elements of the system.**

resemblance to a tower-top focus collector have been investigated by various scientists and engineers who were interested in developing high power thermal sources using solar technology.

7.5.2.1 Alternate Design Approach for a Heliostat System

An alternate approach involving linear concentrator solar arrays has been proposed by several other investigators. One of the investigators [5] has proposed a unique design concept, where a solar energy is concentrated on a metallic pipe inside an evacuated glass tube which is placed at the focus of a cylindrical mirror or lens. A greenhouse effect is produced within the metallic pipe whose surface is coated with selective films capable of absorbing solar radiation with high efficiency. The heat generated on the pipe surface is removed by a circulating fluid using either a pump or heat pipe. The heat is stored as latent heat in a reservoir of salts with high heat-retaining capabilities. Concentration ratios greater than 100:1 have been achieved from these proposed systems, which result in operating temperatures in the 600 to 750 K range (i.e., 327°C to 477°C). It is important to mention that radiation losses, convective heat losses, and fluid pumping losses will limit the plant size for cost-effective operation.

7.5.3 Major Benefits of Tower Top Focus Collector Systems

The tower top focus collector system avoids much of the heat-transport problem associated with linear arrays by transporting the heat energy to a central point in the form of light. In addition, high concentration ratios are possible with the point-focus concept, which permits operating temperatures exceeding 1000 K or 727°C. At such temperatures moisture-free superheated steam can be generated, which is most ideal for operating turbo-alternators with large electrical power capability. Only beam radiation is collected with the focus collector system.

7.5.4 Impact of Critical Element Parameters on System Performance

Mirrors play a critical role in determining the system performance. The aiming accuracy is the most important parameter of the mirror. The mirror surfaces can be made either from silver-plated glass or front-surfaced aluminum plate with a protective coating. Each mirror must be mounted in a rigid support frame to form a heliostat, because frame rigidity is the principal requirement for the heliostat. The angle subtended by the solar disk at the earth is typically 0.5°, whereas the boiler presents twice this angle or 1° for the thickest mirror. The RMS or one-sigma aiming accuracy for the heliostat and control device must not exceed 0.25° (one-sigma value) to avoid excessive beam spreading from flat mirrors. Excessive beam spreading will degrade the thermal performance of the system. Accuracy requirements for the mirrors depend on heliostat dimensions and mirror types.

For a 10 m² heliostat, parabolic mirror segments must be designed with surface accuracy of 1 mm. If the mirror offers a perfect parabolic segment, the image subtends an arc of 0.5°, which will produce an image size of about 9.4 m at the focal point. The mirror flat tolerance gives an indication of rigidity required in the mirror support frame to reduce the defects due to gravity load and wind stress [5]. Tight mirror tolerances are required for radiation temperatures exceeding 1000 K or 727°C.

7.5.5 Impact of Environmental Effects on Mirror Surface

Besides stringent mirror tolerances, the mirror surface must be protected from the adverse effects of the environments. In order to avoid deterioration of the mirror surfaces from sand, dust, snow, and rain when inoperative, appropriate protection techniques must be used. Electrostatic methods must be investigated for keeping the mirror surface clean. Even though the heliostat structure is designed to withstand winds ranging from 120 to 160 km/hour, serious surface deterioration can be expected even in wind at only 10 to 15 km/hour, particularly in desert environments.

7.5.5.1 Performance Parameters of Critical Elements of the System

Detailed calculations of the energy collection efficiency of the mirror field and the boiler are very difficult due to various design parameters and environmental factors involved. The redirected power, which is defined as the solar power directed toward the boiler, assumes lossless reflection from the aberration-free mirrors and this assumption might affect the collection efficiency of the system. As a matter of fact both the effects of mirror aberrations and other losses must be considered, when the design parameters and boiler collection efficiency are discussed.

A preliminary design configuration of a mirror field can be considered if a completely empirical approach is acceptable. It should be assumed that the heliostats will be permanently located and steered to direct the solar image or radiation to a central tower. Snell's law requires the mirror to be positioned at an angle θ, which is defined as

$$\theta = 1/2 \; [\gamma + \beta] \tag{7.1}$$

where γ is the fixed angle from a given mirror to the boiler as shown in Figure 7.9, and β is a variable angle from the mirror to the sun. Under these geometrical conditions, one edge of the mirror will rise above the center and will also change the area of the ground in the shade of the given mirror with respect to the position of the sun and with respect to the boiler location. It is important to mention that any part of another mirror placed in such a "shaded" region will be of no use in producing the redirected solar power.

7.5.6 Preliminary Design Approach

The preliminary or the first design approach for the system is to select a specific season, preferably midwinter, and then to draft a linear north-south array of polished mirrors, which should not experience shading and that would still intercept as large a fraction of the incident solar power as is possible. Under these assumptions, the system design gives the fraction of the total incident solar power redirected regardless of the season in the year. Performance of the mirror is dependent on the redirected fractional ratio from glass to ground (F_{ew}) in the array. The east-west array configuration yields no shading at a specific hour which happens to be 9:00 am. The same array can work in the afternoon as well, if the array can be made symmetric about the tower by increasing the separation of the heliostats on the side nearer to the sun at 9:00 am. This arrangement will reduce significantly the number of heliostats required that will not only reduce the system cost, but also will bypass a small fraction of the solar power. Under this array configuration, system operations at 9:00 am and 3:00 pm are equivalent. Preliminary array performance evaluation indicates that a larger fraction of the solar power is bypassed near noon, while further from noon nearly 100 percent of the solar power incident on this linear array will be redirected and can be used efficiently.

7.5.6.1 Estimation of the Power Redirected by the Mirrors

The net solar power redirected can be determined for each hour of the day in various seasons. Surface reflection and absorption losses must be taken into account, if precise estimation of the redirected solar power is the principal objective. The net power redirected for each hour of the day can be determined using both the east-west (E-W) and north south (N-S) area-based fractional ratios. Expression for the combined fractional factor can be written as

$$F = [F_{NS}][F_{EW}] \tag{7.2}$$

where F_{NS} and F_{EW} indicate the north-south and east-west fractional ratios from glass to ground, respectively. Typical fractional values of glass to ground for N-S and E-W are 0.617 and 0.725, respectively [5]. Inserting these values in Equation (7.2), the value of parameter F comes to 0.4473.

The total mirror area required can be written as

$$A_{mirror} = [F][A_L] \tag{7.3}$$

where A_L is the land area of the concentrator element.

It is important to mention that this mirror configuration offers slightly degraded mirror performance, because diagonal shadows will miss adjacent mirrors in both the N-S and E-W arrays. A computer optimization is necessary to include the diagonal shadows to achieve optimum performance of the mirror.

7.5.6.2 Techniques to Achieve Optimum System and Mirror Performance

Improved system performance is contingent on the optimum performance of the mirror. The mirror optimization factor f_m, which is defined as the ratio of the energy reflected to the tower by the mirror to the solar energy incident upon an equal area of the level ground. This optimization factor will change as a function of how level the ground is. If the mirrors are too close to each other, this factor could be less than one, because of the "shading" effect of either the sun or the tower. For an isolated mirror, the optimization factor could be greater or less than one, depending on the angle of sun above the horizon, which is defined as β [see Equation (7.1)] and the orientation of the mirror. The parameter f_m represents the average value for the entire array and must be determined graphically or by computer analysis. In practice, the competition of these two effects and the economic factors will determine the mirror spacing, and consequently the parameter F. For a specific case optimized for low sun angles which normally occur in the winter season, the mirror optimization factor f_m varies from 0.78 in midsummer to 2.0 on a winter afternoon. For a very rough fit case, the upper value of this parameter can be empirically obtained using the following expression:

$$f_m = [0.780 + 1.5 (1 - 2\beta/\pi)^2] \tag{7.4}$$

7.5.6.3 Performance Parameters for the Boiler and Solar Collector

Seasonal beam energy incident on level ground and beam energy redirected by the mirrors play key roles in determining the performance of the boiler. Computer simulation is necessary to develop curves showing the power redirected to the boiler per square meter of the mirror surface as a function of the time of day and various seasons involved, by incorporating the mirror utilization factor f_m. In 1973, professors at the University of Houston developed such curves, which are shown in Figure 7.10 [5], illustrating the configuration of the mirror field and location of the tower, along with curves for different seasons. Figure 7.10 shows curves indicating the solar radiation intensity at ground level in midwinter. Examination of these curves indicates that winter operation is somewhat improved, but at the expense of peak summer performance of the system. It is quite evident that these curves are relatively flat due to performance optimization over three hours from noon.

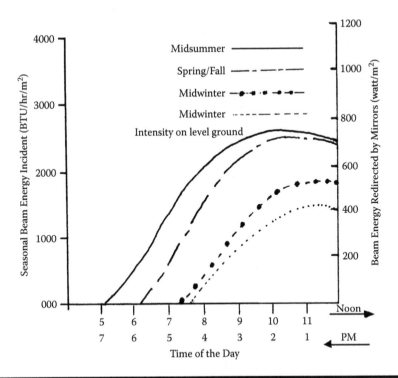

Figure 7.10 **Beam energy incident on ground level and energy redirected by mirror. Note 1 kW = 3413 BTU/hour; area assumed to be 15 h (m²) where h is the boiler height above the ground.**

Boiler performance evaluation is very complex because it involves some assumptions and actual values of several parameters. Total solar power available, radiation losses from the boiler, and transport losses occurring in the pipes carrying the steam must be considered to determine the overall performance of the boiler. The expression for the net power available from the solar system, including the radiation losses can be written as

$$P_{net} = [\text{Solar power incident on the collector}] - [\text{Power losses due to radiation}]$$
$$- [\text{Power losses during transportation}] \tag{7.5}$$

The above expression for the net power available can be written as

$$P = [E_{inc} f_m \, F A_L \, \eta \, \alpha] - [T^4 A_r \, \sigma \, \varepsilon] - [L_{trans}]$$

where L_{trans} is the power loss during transportation, E_{inc} is the incident solar energy, A_L is the land area of the concentrator element, the product of FA_L is the mirror area, f_m is the mirror utilization factor, α is the boiler black body absorption

Table 7.1 Seasonal Beam Energy Available from the Mirror as a Function of Incident Solar Energy in Midwinter

Solar Energy Incident on Ground		Beam Energy Redirected by Mirrors (W/m²)		
(BTU/h/m²)	(MW)	Midsummer	Fall	Midwinter
500	0.33	150 (6:00 am)	150 (7:00 am)	150 (8:00 am)
1000	0.67	275 (7:00 am)	275 (8:00 am)	275 (9:00 am)
1500	1.00	425 (7:00 am)	425 (8:00 am)	425 (9:00 am)
2000	1.34	540 (8:00 am)	540 (9:00 am)	540 (Noon)
2500	1.65	715 (9:00 am)	715 (10:00 am)	525 (Noon)

coefficient, which has a typical value of 0.9, η is the fraction of the solar energy striking the mirror that arrives at the boiler after absorption and reflection by the mirrors are accounted for, T is the temperature in degrees Kelvin (K), A_r is the radiating area of the receiver, ε is the emissivity, which can be assumed 0.9, and σ is the Stefan-Boltzmann constant of 5.67×10^{-8} W/m² deg K⁴. Note the transport losses occur during the pumping of the liquid and heat losses occur in carrying the solar energy collected by the mirrors.

7.5.6.3.1 Computation of Power from Incident Solar Energy

Inserting the assumed values for the above parameters and a maximum value of 2 for the midwinter afternoon mirror utilization factor in the first quantity of Equation (7.6), one gets the computed values of seasonal beam energy incident on level ground, as summarized in Table 7.1.

The values shown in Table 7.1 have been computed using the curves for three seasons, as shown in Figure 7.10 and assuming an area equal to 15.12 h^2. The parameter h stands for height of the boiler above the ground and a value of 500 m is assumed for these calculations. Furthermore, computed values of beam energy redirected by mirrors do not include the effects of surface reflection and absorption losses.

It is important to mention that both the seasonal beam energy incident and the beam energy redirected by the mirrors are strictly dependent on the time of the day and the season involved. Both the beam energy incident and the beam energy redirected by the mirrors occur between 9:00 am and noon, regardless of the season. However, maximum values of these parameters are only possible during the midsummer regardless of the time of the day. These values are based on the mirror area equal to 15.1 h^2 where h stands for the tower height above the ground. Typical tower height varies between 400 and 500 ft depending on the terrain and wind conditions.

Table 7.2 Computed Values of Radiation Losses as a Function of Temperature

Temperature (K)	Radiation Loss (W)
250	199
500	3,189
750	16,150
1,000	51,030

7.5.6.3.2 Radiation Loss Computations

The radiation loss is strictly dependent on the mirror area, temperature, emissivity, and the black body absorption coefficient. Expression for the radiation loss can be written as

$$L_{rad} = [T^4 \, A_{rec} \, \varepsilon \, \sigma] \text{ Watt} \tag{7.6}$$

Inserting the value of 0.9 for the black body absorption coefficient, 0.9 for the emissivity, 1 m^2 for the receiver area, and 5.67 × 10^{-8} for the Stefan-Boltzmann constant, radiation losses have been computed, which are summarized in Table 7.2. Note these radiation losses are insignificant compared to the solar energy incident on the ground, as shown in the second column of Table 7.1.

Transportation losses are very difficult to compute. These computations require physical parameters of the pipes, thermal properties and insulation thickness used for the pipes, boiler dimensions and the steam characteristics such as temperature of the fluid or steam and the entropy.

7.5.6.3.3 Efficiencies of Various System Elements

The overall system efficiency is dependent on the efficiencies of various elements, such as the solar collector, boiler, and mirror. The practical cycle efficiency (η_p) and the Carnot cycle efficiency (η_c) play critical roles in determining the efficiency of the system. Efficiency of the system is the product of solar collector efficiency and the practical cycle efficiency, which is typically 70 percent of the Carnot cycle efficiency. The Carnot cycle efficiency is defined as

$$\eta_c = [(T - 323)/T] \tag{7.7}$$

Computed values of Carnot cycle efficiency and practical cycle efficiency are summarized in Table 7.3. Curves plotted for Carnot cycle efficiency (E_c) and practical cycle efficiency are displayed in Figure 7.11. The computations reveal that both the efficiencies increase with increase in temperature.

Table 7.3 Computed Values of Carnet Cycle and Practical Cycle Efficiency as a Function of Temperature (Percent)

Temperature (K)	Carnot Cycle Efficiency (%)	Practical Cycle Efficiency (%)
323	0	0.003
400	0.193	0.135
600	0.462	0.323
800	0.596	0.417
1,000	0.677	0.474
1,200	0.731	0.512
1,400	0.769	0.538
1,600	0.798	0.559
1,800	0.820	0.574

The overall system efficiency is the product of the solar collector efficiency and the practical cycle efficiency. The collector concentration ratio, designated C in Figure 7.12, is for each concave product efficiency curve at the vortex formed by the two respective curves represented by solid lines and dotted lines. It is important to point out that the transportation losses, surface reflection losses, radiation losses, and absorption losses have been neglected, because they are very small compared to the solar intensities shown in Table 7.1.

7.5.7 Economic Feasibility of the Tower Top Focus Collector System

Comprehensive economic feasibility study is essential to determine the cost-effectiveness of using a tower top focus collector system to generate electrical power. This system involves precision focusing mirrors, a boiler mounted tower with height exceeding 1000 ft, a black body solar collector system placed in the focal region, and a turbo-alternator capable of generating AC power. Furthermore, a large array comprising flat mirrors is required with automatic steering capability to redirect the incident solar radiation to a central point. Furthermore, it is necessary to space the mirrors precisely and to have the collection point elevated high above the mirrors to minimize shading of one mirror by another. It is absolutely necessary to keep the mirror surfaces clean regardless of the operating environments to maintain the aiming accuracy better than 0.25° for the heliostat. All these critical design and stringent operating requirements will increase the system installation costs. Cost-effective system design will require economic feasibility studies for the critical

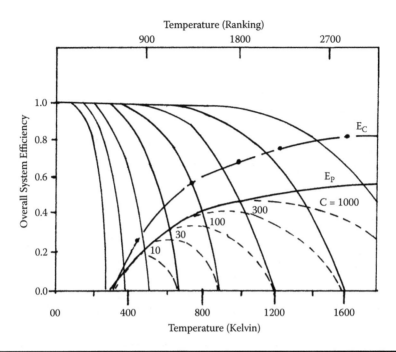

Figure 7.11 **System efficiency as a function of temperature, which is the product of solar collector efficiency and E_p. E_c, Carnot cycle efficiency; E_p, practical cycle efficiency, which is typically equal to 70 percent of E_c; C, collector concentration ratio, which represents the concave product efficiency curve (dotted) at the vortex formed by the two respective curves. The concentration and the system efficiency must be expressed multiplied by 0.794.**

components of the system. Preliminary studies undertaken by professors at the University of Houston (Texas) during 1973 indicate that cost per kilowatt-hour thermal approaches a comparable price using oil or gas as fuel. Since these studies were undertaken about 35 years back, it will be essential to perform cost analysis using present prices for gas and oil to evaluate the current economic feasibility of this approach.

The economics of the tower focus solar energy collector system depend strictly on the mass production of high-quality heliostats with minimum cost and built-in maintenance provisions, if possible. Scientists and design engineers believe that a 50-m² heliostat can be mass produced using state-of-the-art production techniques and low-cost materials for $3000 to $4000. The cost of maintenance of the mirror field could be about 15 percent of the capital cost for the operating life of 30 years. The tower is a relatively straightforward structure, which can be slip-cast of reinforced concrete. The estimated cost for a tower structure could range from $10 to $19 million. The boiler and a sodium-to-steam heat exchanger could present a unique problem, namely, absorbing the heat generated at a flux density as large as

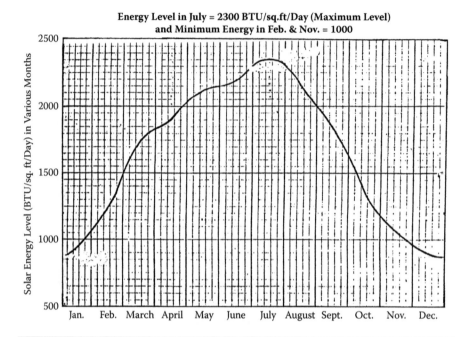

Figure 7.12 Solar energy levels in various months in a year (BTU/sq ft/day).

1.6 MW/m² due to larger concentration ratios. The cost of these two items is a small fraction of the tower cost.

An exact cost estimation for each critical element of a tower focus solar energy collector system is impossible, because the procurement costs are dependent on market conditions and number of items to be bought. The overall cost estimate for such a collector system include installation costs for the tower, receiver, heat exchangers and mirrors, local taxes and interest, operational cost for facilities, auxiliary item costs, and maintenance costs.

According to professors at the University of Houston [7] rough estimates for various system elements can be summarized as follows:

■ Procurement costs for 500-m tower, receiver and heat exchangers: $25,000,000
■ 2.5 km² mirror field with 45 percent coverage at $50/m²: $56,250
■ Total capital investment required: $25,056,250
■ Depreciation based on straight line rule for 20 years: $1,252,812
■ Local taxes and interest at 8 percent: $2,004,500
■ Maintenance for mirror field, 25 men at $30,000/year: $750,000
■ Manpower required to maintain tower facilities, 5 men at $20,000/year: $100,000
■ Materials, supervision, and overhead costs (estimate): $500,000
■ Estimated net system cost excluding return on investment: $4,606,312

The return on investment will depend on net system cost, net cost per kilowatt/hour-thermal, cost of natural gas per thousand cubic feet, and equivalent cost per barrel of oil (one barrel produces 6 million BTU heat energy or 1758 kW of electrical energy). Comparable electrical generation cost per kilowatt-hour depends on whether one uses natural gas or oil or solar cell technology.

7.5.8 Impact of Solar Energy Levels on the Tower Focus Solar Energy Collector

The thermal output of the tower focus solar energy collector system, which is defined as thermal energy expressed as BTU/sq ft/day, is strictly dependent on the solar energy level available in a particular month of the year. The curve shown in Figure 7.12 illustrates the solar energy available in various months of the year in the northern hemisphere. This curve indicates that in the northern hemisphere, maximum solar energy level is available in the month of July, and minimum solar energy levels in the months of January, February, November, and December. The solar energy levels will be in reverse order for tower focus solar energy systems operating in the southern hemisphere. Countries located near the equator will see higher solar intensity levels compared to those located near the Arctic Circle region. Furthermore, one could expect desert regions to have relatively higher solar energy levels regardless of the month in the year. Based on these statements, tower focus solar energy collector systems will be found most cost effective and beneficial when operated in tropical countries located near the equatorial regions. As stated earlier that maximum solar intensity occurs during midsummer regardless of the time of the day. However, optimum energy generation from a tower top focus solar energy collector system is possible between 9:00 am and 3:00 pm on sunny days. One can visualize system benefits from the curves shown in Figure 7.11, where relatively flat regions seem to offer optimum system performance over a three-hour time frame on either side from noon. Tower height is a critical system parameter, because it affects the system structure cost, mechanical stability of the boiler, pumping of liquid for heat transfer, and maintenance procedures. In summary, serious considerations must be given to design complexity and capital investment cost, before making a decision to install a tower top solar energy collector system.

7.6 Summary

This chapter describes the design and operational aspects of grid-connected and stand-alone photovoltaic (PV) power generating systems. Major advantages and disadvantages of two PV systems are briefly summarized. Stand-alone PV power systems include communication and navigation satellites, radio relay stations, unprotected railway crossings, educational TV broadcast facilities for remote areas, marine batteries, electronic watches and portable calculators. Major potential users of grid-connected PV power systems are residential housings, commercial buildings, manufacturing facilities, and discount stores such as WalMart, Target, or KMart. The solar panel is the critical element of a grid-connected solar power system. Design requirements and performance parameters for roof-mounted solar panels are discussed, with emphasis on cost, reliability and sustained power output over the life of the panel. Performance capabilities and limitations of solar panels are summarized.

Major advantages of a stand-alone solar power system are identified. Design aspects and performance capabilities of open-loop and closed-loop water heaters are briefly discussed, with emphasis on cost and performance. Component requirements and performance parameters for a closed-loop water heater are summarized, with emphasis on economic benefits. Design configuration and critical component requirements for a low-cost domestic water heater are discussed. Design requirements and cost estimates for solar-based hot water are summarized to meet hot water consumption requirements for a family of four. Solar panel design configuration and parameters for a pool heater system are identified, with emphasis on cost and reliability.

Advantages and performance capabilities of a tower top focus solar energy (TTFSE) collector system are summarized, with emphasis on overall system performance and complexity. Critical design parameters and operational requirements for this system are briefly discussed, with emphasis on system cost and complexity. Design requirements of heliostats best suited for a TTFSE system are defined. Impact of critical performance parameters on the overall system performance are identified, with emphasis on cost and reliability. Computations of redirected power by mirrors are provided for the benefit of system designers. Design and operational aspects of a tower-mounted boiler are discussed, with emphasis on transportation losses caused by the circulating fluids. Computations of redirected beam energy by the mirror as a function of solar energy incident on the ground, radiation losses, and transportation losses are provided to identify the potential benefits of the system. Data on economic feasibility are provided to justify the use of the TTFSE system. A cost breakdown summary for a TTFSE power system is provided to identify high-cost system components.

References

1. Peter E. Glaser. "An overview of the solar power satellite option," *IEEE Transactions on MTT* 40, No. 6 (1992): 123–127.
2. A.R. Jha. "Technical proposal and cost estimation for a solar power system using solar technology." Unpublished paper. Jha Technical Consulting Services, Cerritos, CA, 1998.

3. A.R. Jha. "Low cost solar systems." Technical Proposal. Jha Technical Consulting Services, Cerritos, CA, 1986.
4. Technical Fliers from King Energy Systems, Irvine, CA.
5. F. Hildebrant and L.L. Vant-Hull. "A tower-top focus solar energy collector," *ASME Solar Energy*, Division, University of Houston, Texas, Mechanical Engineering, September 1974, 22–26.

Chapter 8

Performance Capabilities and Economic Benefits of Potential Alternate Energy Sources

8.1 Introduction

The principal objective of this chapter is to describe the performance capabilities and economic benefits of potential energy sources, with particular emphasis on electricity-generating cost per kilowatt-hour, greenhouse effects, longevity, and reliability. Potential energy sources are described briefly, with emphasis on installation cost, design complexity, and uninterrupted availability of electrical energy at reasonable costs. Unique techniques will be discussed for installation of solar panels with minimum cost and complexity, but no compromise in system performance. Requirements for solar panel installation on flat roofs, inclined roofs, and inverted V-shaped roofs [1] will be defined to achieve maximum efficiency of the solar modules with minimum shadowing effects. Economic aspects of energy technology will be discussed, with emphasis on material cost, installation cost, maintenance cost, and initial capital investment requirements. Finally, solar energy conversion cost per kilowatt-hour will be compared with energy generation costs using diesel, coal, natural gas, wind turbine, hydraulic turbine, biofuel, geothermal technology, and nuclear fuel. Harnessing of electrical energy from sea waves and river flows will be summarized, with emphasis on cost, continuous availability of electrical energy, and reliability. Critical issues such as energy conversion, storage requirements, low-cost alternate energy sources, and energy harvesting technology will be briefly addressed with emphasis on affordability. Since solar cell technology is receiving the greatest attention, affordability of solar cell technology will be compared with other energy-generating schemes using steam, coal, oil, and geothermal technology. Performance capabilities and limitations for various alternate energy-generating sources will be summarized, with emphasis on affordability, reliability, and performance.

Studies indicate that the initial investment required for a large hydroelectric power generating plant could be lower than a coal-fired steam turbine-generating plant. In addition, the maintenance and shut-down costs for a hydroelectric power plant are much lower than that for a coal-fired steam turbine-generating power plant. Based on these costs, the cost for electricity from hydroelectric power plants is much lower compared to steam power plants or nuclear power plants.

A hydroelectric power plant requires continuous flow of water from a lake or a large water reservoir located at a higher level than the hydroturbine and several water controlling gates, each assigned to its hydroturbine generator set. The stored water input is first converted into kinetic energy using high-velocity open air jets, which drive the hydroturbine generators. The open air jet velocity is dependent on the height of the water reservoir and the physical dimensions of the controlling gate. The Hoover hydroelectric plant in Nevada presents a classic example of a large hydroelectric power installation.

8.2 Alternate Energy Sources and Their Installation Costs and Electrical Power Generating Capacities

Preliminary studies by energy planners indicate that geothermal sources, solar cells, wind turbines, hydroturbines, and micro-hydro installations are considered the most environmentally friendly renewable energy sources [1]. However, these sources alone cannot meet worldwide energy requirements which grew by 5 percent approximately to 449×10^{15} BTU or 131×10^{12} kW. Petroleum makes up 34 percent of energy sources worldwide, coal 32 percent, natural gas 21 percent, nuclear power 6.5 percent, and renewable sources (wind, solar, geothermal, hydroelectric, tidal, etc.) 6.5 percent [2].

8.3 Energy Sources Best Suited for Various Organizations

Alternate energy sources are available for civilian building and military or defense installations, which would provide electrical energy at relatively lower costs.

Table 8.1 Costs for System Installation and Electrical Energy Generation, and Performance of Various Energy Sources

Energy Source	Typical Efficiency (%)	Installation Cost (Before Rebates)	Generation Cost (cents/kW-h)
Solar cell technology	14–18	$32,000 (5 kW)	12–20
Wind turbine	65–75	$10–$12 million (100 MW)	5–8
Hydroelectric	55–68	$100–$200 million (1 GW)	10–18
Steam turbine	45–65	$80–$100 million (1 GW)	8–15
Nuclear reactor	50–70	$6–$8 billion (4 GW)	6–12
Ocean tidal wave turbine	38–55	$5–$8 million (200 MW)	15–18

University buildings, commercial buildings, shopping centers, and defense installations or military bases must seriously consider alternate reliable energy sources to ensure modest supply of electricity with lower electricity costs and free from generating greenhouse effects. In fact, a hybrid approach consisting of solar cell and wind turbine technologies may be a cost-effective energy source, which could meet electricity requirements for the base schools and military personnel living quarters. However, a total solar cell approach will be a more cost-effective and reliable electrical energy source. Geologists can perform soil analysis for various locations on the base and recommend the right operating location to tap the geothermal energy source. These energy sources will not only realize significant reduction in energy cost, but also minimize the dependence on foreign oil, which should be of major consideration during times of international conflicts.

Pertinent data collected by the author on design complexity, efficiency, installation cost, and cost of electricity generation from various energy sources [2] are summarized in Table 8.1.

The accuracy of the tabulated data is not guaranteed, because they have been quoted from various sources. However, they are likely accurate within ±20 percent. Furthermore, the installation cost is dependent on several factors, such as power generating capacity, construction cost, stringent safety requirements, type of fuel used, type of turbine and its mechanical aspects, plant installation locations, and electric energy interface requirements. The values shown in parentheses indicate the electrical energy generation capacity of the plant.

Date presented in Table 8.1 indicates that solar cell power systems, wind turbines and tidal wave turbines require minimum capital investments. As far as the electricity generation cost is concerned, wind turbines and nuclear reactors appear to be the most ideal alternate energy sources. As far as energy-generating capacity is concerned, nuclear reactors offer the highest energy-generating capacity, exceeding 5 GW. However, nuclear power installation cost could be as high as eight billion dollars. Installation costs for hydroelectric turbine- and steam turbine-based power-generating systems could be $100 million or more depending on the power ratings of the turbines and alternators. Note the installation cost projections or estimates for wind turbines and tidal wave turbines could have errors of ±10 percent. However, installation cost projections for nuclear and hydroelectric power plants could have errors greater than ±20 percent due to the large number of assumptions and variables involved. As far as construction durations are concerned, nuclear, hydroelectric, and steam power-generating plants could take from two to four years depending on the completion of other associated components or subsystems involved. In summary, data presented on installation cost estimates and electrical energy generation cost show basic trends.

8.3.1 Geothermal Energy Source

According to the studies performed by geologists, geothermal power plants with 250-MW generating capacity are possible in desert areas. Such a geothermal power plant can meet more than 10 percent of electricity requirements for military bases, where ample lands are available for solar panel installations. Geothermal power plants generate the electricity through wells, which can be a thousand feet deep into the earth. These wells bring heat to the surface by drawing hot water or steam generated from the high-temperature cracks in the earth. The hot water can be used to heat the base buildings, quarters for military personnel, and university research laboratories. The steam can be used to drive a turbine-generator set to generate electricity. This steam is not superheated and is only suitable to drive low-speed turbine-generator sets.

Finding the best spots to drill deep wells can be tricky as well as expensive. Geothermal electric power plants require the services of experienced geologists to find out the most suitable locations for geothermal power plant installation, which could be costly. A quick scan through Google Earth images can reveal pressurized linear features that could turn out to be faulty. Selection of the right location requires analysis of the rocks, examination of soil densities, and comprehensive evaluation of characteristics of the subsurface water, in addition to magnetic and gravitational field measurements of the site selected. A sudden and unusual magnetometer reading suggests the presence of a flowing fluid in the rock, which points to a well-defined location for the geothermal energy source [3]. However, experimentation using "LIDAR" laser imagery would add more certainty to the geothermal site selection. In brief, finding the right location for the geothermal energy source [3] is time consuming and costly, but this energy source has no greenhouse effects and is independent of foreign oil.

8.3.2 *Solar Power Installations*

Solar power installations are best suited for utility companies, large department stores, and military installations. Solar panels require large flat and inclined roofs to get maximum benefit from the incident solar energy. If no electrical utility is available nearby, a hybrid electrical power system comprising a wind turbine and diesel generator plant is the best alternate electrical energy source. Coastal regions are best suited for installation of wind turbines. For example, military bases have lots open lands, where solar panels can be installed with minimum cost and complexity to provide an alternate source of electricity for the base schools, offices, and personnel quarters. In remote military installations, backup diesel generators must be used to supply electricity to critical equipment and buildings, in case the solar power installation is unable to meet the base electrical power consumption requirements.

A solar power installation employing photovoltaic (PV) arrays in the desert of Las Vegas, Nevada, is operating to meet electrical energy requirements exceeding 15 MW. Hundreds of half-horsepower motors are deployed to adjust the inclination angle for the 72,416 solar panels to the west for optimum system performance during the day. As dusk approaches, the solar panels are required to tilt backward to the east. According to the solar panel installers, if the panels turn westward as dusk approaches, the shadows cast by some panels onto the others will lead to greater efficiency loss than if the solar cells soaked up solar rays while lying flat, at an oblique angle to the descending sun. The fine movements of solar panels are computer-controlled tracking mechanisms. The solar panels, initially fixed at a 20° southward tilt, rotate from east to west in response to the tracking command signals. Each string consists of a few dozen solar panels with its own motor and a controller. The controller feeds the data to an inverter, which is remotely monitored in the control room. On a hillier section of the solar panel installation, the strings are programmed to move along paths that take into account the terrain features. Weather stations throughout the installation site are required to record the air temperatures, wind speeds, and solar radiation intensities to monitor solar system performance. A monitoring algorithm uses these atmospheric parameters to calculate how much electricity the solar power system should be generating at any given time. Based on the performance records of solar systems, tilting mechanisms and tracking algorithms allow the solar installation to generate roughly 30 percent more electrical energy than if the solar panels were fixed to one position.

Precise determination of capital investment and breakeven point for a 15-MW solar power installation is extremely difficult. Based on the available published data for solar installations exceeding 15 MW, the cost of those trackers, inverters, solar modules, and concrete pedestals add up to more than $100 million and breakeven point on such a solar system could be approximately 50 years.

Regarding the electrical power-generating cost from a solar power plant, again actual generating cost per kilowatt-hour is dependent on several factors, such as solar panel cost, installation cost, concrete pedestal erection costs, number of pedestals required, number of tracking mechanisms and monitoring algorithms involved, and weather monitoring stations needed to record the atmospheric parameters. A 15-MW solar installation offers electricity at 3 cents per kilowatt-hour to a military base under specific contract obligations and renewal-energy credit incentives. In another case, an Air Force base receives electricity from a solar installation at a higher price of 13 cents per kilowatt-hour compared with a national average of 9 cents available from the local utility company. Despite these prevailing rates from the solar energy power-generating sources, military bases plan to produce more electricity from solar sources on site than they consume from the utility grids or from other fossil energy sources.

8.4 Hydroelectric Power Plants

Large amounts of electrical power can be generated from hydroelectric turbine generators with lower costs than from coal-fired steam-turbine generators or diesel power-generating sets. Furthermore, both the operating and control rooms enjoy cooler and safer environments compared to steam-turbine power plants. No energy source is greener than hydroelectric power-generating plants. Hydroelectric power plants similar to the Hoover Dam Power Installation require billions of dollars for erection of the plant and could take three to five years for completion. Furthermore, an enormous supply of dam or reservoir water is essential for operation of the hydroelectric turbines. The electrical power-generating cost is slightly less than the energy generated by a coal-fired steam turbo-generator plant, because of its lower operating and maintenance costs. In addition, the operational life of a hydroelectric power plant is higher compared to steam-turbine or nuclear power plants.

8.4.1 Micro-Hydroelectric Power Plants

Hydroelectric plant installers claim that construction and operating costs for micro-hydroelectric plants would be lower and do not require large capital investment and long completion period. Small hydroelectric plants, known as micro-hydroelectric plants, require minimum artificial impoundment and produce the lowest levels of greenhouse gases. Even with adequate water source, efficient and cost-effective implementation of hydroelectric turbine generation and associated control systems could pose some practical problems.

One of the biggest challenges facing the designer of a micro-hydroelectric control system is how to accommodate variable electrical loads while operating the hydroelectric turbine at its constant maximum-power specific speed. The control system circuitry provides this function by automatically adjusting the alternator field excitation current with output voltage feedback. Conversion efficiency is the most critical performance parameter of any electrical power-generating system. In the case of a hydroelectric power plant, efficient generation of electricity depends strictly on a good interface between the hydrodynamics of the water-driven turbine and its electromechanical load. A variety of turbines can be employed in micro-hydroelectric installations, which include "impulse" turbines and "reaction" turbines. Impulse turbines are widely used because of the relative simplicity of their unpressurized housing, leading to lower installation costs.

In impulse hydro-turbines, the hydrostatic energy, which is a product of volume and pressure of the input water, is first converted into kinetic energy (mass × velocity^2) by one or more high-velocity open-air jets, which in turn impinge upon and drive the hydro-turbine generator set. An impulse turbine is characterized by a specific speed, which is defined as the speed of rotation (RPM) for a given jet velocity, that provides the most efficient conversion of water energy to mechanical energy. The jet velocity produces maximum output for any given hydrostatic head input. The water jet velocity is determined by the water source pressure or hydrostatic head, which can be expressed as

$$\text{Hydrostatic head} = [P/500]^{1/2} \qquad (8.1)$$

where P is the source or water pressure and is generally expressed in lb/sq ft. Therefore, the hydrostatic head will be given in feet. Based on this equation, the impulse turbine achieves the maximum power-point (MPP) tracking implicitly, without the need for a separate tachometer or other appropriate means to monitor the turbine speed. The output voltage equation based on the theory of an alternator can be written as

$$V_{\text{out}} = [K\,(RPM)\,I_x] \qquad (8.2)$$

where K is the positive feedback factor, RPM is the rotational speed of the turbine, and I_x is the excitation current from the exciter, which is coupled directly to the alternator [4]. If the feedback factor K is manually adjusted so that the resulting constant RPM is equal to the maximum-power specific need of the turbine, the MMP tracking will be automatically achieved. Since, the feedback inherently comprises positive feedback, chances are that the turbine will become unstable and can oscillate violently without ever converging to a stable operating point [4]. To overcome this instability problem an appropriate algorithm or a novel integrating temperature-controlled algorithm must be used. Studies performed by the leading turbine designers reveal that the microhydro-turbine controller can maintain the turbine's maximum-power point (MPP) speed, while using a "take-back-half" feedback algorithm to eliminate the possibility of instability.

8.4.2 Benefits of a Microhydro-Turbine Generator

A microhydro-turbine generator offers more benefits compared to a conventional hydroelectric-turbine generator system. The major benefit of a microhydro-turbine is the initial installation cost. Microhydro-turbine installation is possible in the vicinity of a man-made water storage tank that is receiving constant water supply from a river, or close to a lake that is getting water supply through the large water pumps located at the output ends of the microhydro-turbines. This water feed scheme permits the required level of water at the main storage area. As stated before, the installation costs for a micro-hydroelectric power plant are much lower than the conventional hydroelectric-generating plants, because of minimum maintenance and operating costs. Furthermore, minimum performance monitoring staff is needed in the control room. It is important to point out that electricity generated from a micro-hydroelectric power plant does not contribute to greenhouse effects.

8.5 Steam Turbo-Alternator Power Plants

Coal-fired steam turbo-alternators (CST) have been in use worldwide since 1920 or even earlier. CST power plants are widely used in countries that have no shortage of coal. The cost of electricity generated by such a plant is strictly dependent on the distance between the CST generating power plant and the coal mines, because all other cost contributing factors remain the same. Therefore, for a reliable and uninterrupted electricity source, the CST must be as close as possible to the coal mines. Ashes from the boilers and smoke from the chimney are not environmentally friendly. The smoke from the chimney contains large amounts of hydrocarbons, which are extremely harmful for the eyes and lungs.

8.5.1 Anatomy of a Steam Turbo-Alternator Power-Generating Plant

It is important to mention that the steam turbo-alternator power-generating (STPG) plant requires a continuous supply of coal for daily use and surplus coal stock is necessary to meet electrical power requirements for 30 to 45 days. High-pressure superheated steam from a boiler is essential to generate electricity with higher conversion efficiency. The boilers are installed on top of furnaces and are equipped with special groups of steel tubes, which generate superheated steam best suited for a high-speed turbine coupled to pinion for reduction in speed compatible with the alternator RPM. The exciter is directly coupled to the alternator, which provides the DC electrical power to produce the magnetic field. The AC electrical power generated by the alternator is sent to high-voltage power transformers for transmission to various distribution centers, where the voltage is reduced to 110 V and 220 V for customer use. After leaving the low-pressure turbine section the steam goes to a condenser, where it is condensed to high-temperature water. This hot water is sent to a cooling tower via circulating pumps and the cold water is sent to boilers. This completes the operating cycle of an STPG power plant. Based on best estimates, the cost of electricity generated by such a power plant varies from 8 to 15 cents, depending on the distance between the power-generating plant and the coal mines. It takes about one to two billion dollars and more than two years to set up an STPG. The cost of electrical energy generated by an STPG power plant is lower compared to solar cell-based energy, which varies from 24 to 30 cents per kilowatt-hour.

8.5.2 Maintenance and Operating Costs for an STPG Power Plant

In steam-based power plants, trained operating personnel are needed on a 24-hour basis to monitor the performance parameters of the boilers such as water level and superheated steam pressure and temperature. At least one person is constantly required to monitor the turbine and alternator performance parameters. Sudden decrease or increase in electrical loads could affect the turbine speed and alternator electrical parameters.

8.6 Nuclear Power Plants

Both the capital investment and the time to plant completion are highest compared to other types of power-generating plants. It takes roughly two to four years and requires three to eight billion dollars for the completion of a nuclear power plant. However, one can generate 4000 to 10,000 MW of electrical power from nuclear power plants, which is not feasible from other power sources, such as wind power, solar-steam, solar Stirling engine, photovoltaic cells, fuel cells, tidal waves, and bio-mass. The initial investment required for the nuclear reactor, thermal and radiation shields, radiation protection requirements for the cooling tower, steam turbo-alternator set, and recycling of spent nuclear fuel are enormous. Extraction and storage of nuclear waste from a nuclear power plant present costly environmental problems. The nuclear waste must be stored in radiation-proof containers and must be buried deep (about 50 to 100 feet) in the ground in isolated locations. Furthermore, selection of a nuclear plant location requires comprehensive survey with emphasis on availability of cooling water supply, remote from populated areas, and close to beaches or rivers for easy access.

8.6.1 Major Design Aspects and Critical Elements of a Nuclear Power Plant

Critical elements of a nuclear power plant include the nuclear reactor core, nuclear fuel rods, steam turbo-alternator, condensing unit, and cooling tower. The plant can be operated over a one- or two-year period without shutdown, unless there is a serious radiation leakage problem or physical danger to the plant from an earthquake. Continuous monitoring of radiation levels is essential in the vicinity of the reactor core, cooling tower, and steam turbo-alternator unit for the safety of the personnel involved. If any maintenance is required, complete shutdown of the plant is necessary and requires measurement of radiation levels in the vicinity of the critical elements of the plant.

8.6.2 Benefits and Drawbacks of the Nuclear Power-Generating Installation

There are advantages and disadvantages of a nuclear power plant. The major advantages include the maximum power-generating capacity and nonstop operating capability over extended durations. The serious disadvantages are high cost of plant installation, nuclear radiation hazard, transportation of nuclear waste, and underground storage of nuclear waste without harm to the environment.

8.6.3 Costs for Erecting the Plant and Electricity Generation

According to the International Atomic Energy Agency (IAEA), based on studies performed during the 1970s, nuclear energy was the energy source of the future. The studies further concluded that thousands of nuclear plants could be built with operating capacities as large as 5000 MW. However, significant escalation between 1985 and 2007 in the cost of installing a nuclear power plant, from one billion to more than eight billion dollars for a power plant with the same operational capacity, has posed serious problems for some developed countries and for most underdeveloped countries. The current cost of a nuclear plant installation for 4000 MW energy capacity varies from six to nine billion dollars, depending on the construction costs and inflation rates in a given country. The cost of building a nuclear plant with capacity exceeding 6000 MW could be significantly higher depending on the cost for the construction materials, radiation-resistance material requirements, level of security involved, and stringent thermal insulating specifications to ensure the protection of the operating personnel.

8.6.4 Reasons for Temporary Setback for Deploying Nuclear Power Plants

Since 1978, more than 18 European countries have voted to approve the phase-out of nuclear power plant installations. Furthermore, nuclear power plant accidents such as those at Three-Mile Island, Chernobyl, and the Japanese Monju reactors have put the brakes on viable options for further increase in nuclear power installations. In addition, the significantly high cost of nuclear power plant construction cannot justify further installation of nuclear plants. The following are the principal reasons for not favoring such power plants:

- Safety of the public and environments
- Problems in disposal and storage of nuclear waste
- Danger of theft and proliferation of nuclear materials by terrorists

■ Relatively high cost of generating electricity (10 to 16 cents perkilowatt-hour) compared to a conventional steam turbo-alternator power plant

8.7 Tidal Wave Energy Sources

Commercial installation of tidal wave or ocean-based energy sources is getting serious consideration. The first commercial ocean-based energy project is scheduled to launch in the summer of 2008 off the coast of Portugal [4]. A Portuguese utility company plans to install wave power-generating modules in the coastal regions. Strong tidal waves are most common in the Portuguese, U.K., and Scottish coastal regions, thereby indicating the best coastal locations to generate electrical energy in great amounts. Britain plans to generate 20 percent of its electrical power using the ocean energy concept. Scotland has a huge renewable-energy potential, enough to meet its demand for power almost 20 times over in a few years. The country has the capability to generate more than 25 percent of Europe's tidal power potential and 10 percent of its wave power potential. New Zealand plans to install 200 tidal wave turbines near Auckland harbor. The turbines will be anchored to the seafloor near the coastal region to generate maximum electrical power. Auckland energy planners expect that these turbines could generate more than 400 MW of electrical energy, which will be roughly 3 percent of the country's energy demand. To accelerate the installations of tidal wave turbines, New Zealand's energy department director has placed a 10-year moratorium on the construction of fossil fuel power plants.

The Portuguese utility company plans to install three snakelike wave-power generators made by a Scottish company, which will generate 2.25 MW of electrical energy. The electricity generated will be carried to coastal towns through a short underground cable. The utility company expects to install an additional 28 such generators, which will boost the capacity to 22.5 MW, enough to meet the electricity requirements for roughly 1800 homes. Countries located in coastal regions with high tidal waves should consider this energy source as a way to meet their renewable energy targets.

8.7.1 Operating Principal of Tidal Wave Energy Sources

The operating principle of this type of power generation involves three distinct schemes: the power of ocean waves [5], the flow of currents, and the motion of the tides to harness the ocean-based power. Wave-energy converters, called linear absorbers, comprise three long canisters that look like giant oxygen tanks. Hinged joints link the canisters. When the ocean waves change the segment's positions with respect to each other, the joints push hydraulic rams. This action will pump high-pressure oil through the miniaturized turbo-alternators, which are located inside the canisters. Rotation of the turbo-alternators will generate AC electrical energy.

8.7.2 Benefits and Drawbacks of Tidal Wave Energy Sources

Strong waves and coastal regions are the principal requirements for tidal wave energy plants. Tidal wave energy sources produce zero greenhouse gases and pose no danger to environments. The tidal-based power plant requires minimum operating personnel. The turbine-generator modules are least expensive compared to those deployed by the coal-fired steam turbo-generator plants or nuclear power plants. Optimum selection of coastal location for anchoring the turbines requires comprehensive survey of coastal regions where strong waves are prevailing. Because the turbines and associated components are continuously exposed to salt water and coastal sand, frequent maintenance procedures may be required to minimize wear and tear of turbine blades and to maintain the turbines operationally fit all the time. Shutdown of the tidal wave power plant is only possible when the turbine module requires maintenance or it is damaged.

8.8 Wind Energy Sources

Wind turbines are widely used in several countries to generate electrical energy to meet the energy needs of the local people. Coastal regions, hilly locations, and open desert lands are best suited for the installation of wind turbines. Significant increase in greenhouse gases has compelled various countries with long coastal regions to consider the potential application of wind turbines to generate electricity. Wind turbines, when installed in coastal regions, yield optimum system performance. According to the Global Wind Energy Council [6] China has installed more wind turbines in 2007 than any other country, and is producing electrical energy exceeding 3 GW (3×10^{12} W). The council further states that China is the third fastest-growing wind turbine market behind the United States and Spain. According to wind energy planners, China was expected to add another 4 to 5 GW of wind turbines by the end of 2008, giving the country a capacity exceeding 10 GW by the end of 2008. Most of the Chinese wind turbines are operating in Inner Mongolia desert regions. These wind energy systems are operating in conjunction with the utility grid system. Denmark is considered a wind-energy pioneer in the entire world.

8.8.1 Affordability and Environmental Benefits of Wind Turbines

Wind turbine installations require open lands, hills, and coastal regions to achieve optimum source of electrical energy. No other energy source could claim greater affordability than wind turbines. The cost of generating electricity is dependent on the wind turbine operating capacity, construction cost, and maintenance cost. Chinese power-generating companies estimate that the cost of electricity generated by wind turbines is close to 6 cents (in U.S. currency) per kilowatt-hour. This indicates that the electricity generated using wind turbines would yield much lower energy cost compared to most other energy sources without polluting the environment. Several countries are seriously considering rapid deployment of wind turbines to take advantage of the lowest electricity generation cost from this particular technology.

8.8.2 Worldwide Deployment of Wind Turbine Technology

Based on the published data [6], Denmark has been recognized as a pioneer in developing the wind turbine technology as early as 1970. In 2003, Denmark built and put into operation the world's largest, off-shore wind turbine with generating capacity of 915 MW. By the year 2006, Denmark had a total wind power installed capacity exceeding 3136 MW, which meets roughly 30 percent of the country's electrical energy needs. Today, 90 percent of the wind turbines are manufactured in Denmark which are sold in international markets capturing 43 percent of the world's wind turbine market. Today, Europe generates about 3.5 percent of its electrical energy using wind turbine technology.

The global wind turbine market grew more than 35 percent by the year 2006, with electricity-generating capacity exceeding 15,300 MW. In 2005, Germany installed a wind turbine with generating capacity of 2300 MW, bringing their total wind turbine installation capacity to more than 21,000 MW.

Britain has added 168 MW to its wind power energy system to bring its capacity to greater than 2100 MW. Poland is planning to add wind turbines, with total generating capacity exceeding 1100 MW by 2010. It is important to mention that Germany is putting a major emphasis on renewable energy programs, which will encourage the deployment of solar cell technology and wind turbines to generate electricity at lower costs.

8.9 Use of Solar Cells to Generate Electricity

Preliminary studies performed by the author on potential nonfossil energy sources such as biomass, photovoltaic (PV), wind turbine, and ocean wave indicate that solar cell technology has zero greenhouse effects and is considered the most environmentally friendly source of generating electrical energy. The studies further indicate that a constant source of generating electrical energy with moderate costs is possible through the deployment of wind turbines and PV technology. Significant reduction in greenhouse effects is the major design consideration in the selection of alternate energy sources.

8.9.1 Estimation of Greenhouse Gas Contents in Various Energy Sources

Estimation of greenhouse gas contents (grams/kilowatt-hour of CO_2) generated by various energy sources can be summarized as follows:

- Coal, 0.900
- Oil, 0.850
- Natural gas, 0.425
- Biomass, 0.450
- Multicrystalline Si, 0.370
- Cadmium telluride, 0.180
- Nuclear reactor, 0.240
- Wind turbine, 0.115
- Ocean tidal waves, 0.800

Based on the above greenhouse gas contents [7] it is evident that solar cells, nuclear power plants, and wind turbines offer the cleanest reliable alternate energy sources. However, nuclear power plants suffer from nuclear radiation hazards and problems of nuclear waste disposal and storage. Wind turbine installations require either coastal regions or vast open lands, thereby making the location of the plant more difficult economically. On the other hand, PV cells can be installed on the roofs of existing homes or commercial buildings, thereby making the solar installation more attractive and affordable. In addition, once the solar panels are installed, no maintenance is required unless they are damaged by earthquake or fire or other natural calamity. The solar panels are guaranteed by the manufacturers for their operating life, which ranges from 25 to 30 years.

8.9.2 Installation and Reliability Requirements for Photovoltaic Cells and Solar Panels

Installation and reliability requirements for the solar cells or photovoltaic cells and panels will be discussed in great detail. Design requirements and performance parameters of various solar cells and solar panels or modules have already been summarized in the preceding chapters. In addition, material requirements and fabrication procedures for solar cells and panels have been discussed in great detail and for specific details, refer to the appropriate chapters. It is important to mention that rapid deployment of solar cell technology would get a grip on global greenhouse effects and would eliminate the dependency on the foreign oil.

8.9.3 Reliability and Operating Life of Solar Cells and Panels

So far very little information has been available on the reliability and longevity of solar cells and panels. There are no serious reliability problems as far as solar cells and panels are concerned. The solar cells or devices are entirely passive and do not require external biasing. Therefore, external bias fluctuations or short-circuit conditions do not exist. As stated earlier, the solar panels or modules are guaranteed for 25 to 30 years by the panel manufacturers. However, the inverter, which converts the DC output energy of the solar panels into AC electrical power to be compatible with the utility grid's electrical specifications, has a life of only 10 years. This is the most critical component of the solar energy collection system. The inverter cost varies from $3000 to $3300 depending on the panel power ratings. Performance of solar cells slightly improves with increase in operating temperature, but operating temperatures up to 400 K pose no reliability problem either to solar cell or solar panels.

8.9.4 Performance Degradation in Solar Cells, Solar Panels, and Inverters

The conversion efficiency and the power output of the solar cells and consequently of the solar panels tend to decrease with the operational life after the installation. According to the solar cell manufacturers, both the conversion efficiency and power output of the solar cells decrease by 20 percent approximately after 25 years of continuous operation. As the solar panels are protected with plastic sheets or glass enclosures, they do not sustain structural damage or deterioration in performance from rain, wind, sand, or snow. Furthermore, the solar panels are designed to provide high mechanical integrity in severe environments including extremes of temperature, vibrations, and humidity. The panel suppliers do not recommend any maintenance requirements for the solar cells, solar panels, and inverters. Furthermore, once the panels are installed on the roof, one cannot have easy or convenient access to the solar cells or panels.

The conversion efficiency of the inverter is the most critical performance parameter in the entire solar power system. The first inverters had efficiencies ranging from 92 to 95 percent. However, according to designers, the efficiency of the second generation of inverters could be as high as 98 percent. Such inverters will cost slightly more, but will have enhanced efficiency, improved reliability, and greater longevity. The large PV generator is coupled to the DC terminal of the inverter, while the utility grid is coupled to the AC side of the inverter. To achieve higher inverter reliability, stringent requirements for harmonic suppression and electromagnetic compatibility (EMC) are essential. Critical design aspects and functional capabilities of the inverter can be summarized as follows:

- Inverter efficiency must be maintained well above 95 percent of the nominal load.
- Stringent harmonic suppression and EMC requirements should be defined.
- Precision and robust maximum power point (MPP) tracking is essential under partial shadowing environments.
- Random data acquisition and monitoring are essential, if the reliability of the inverter is the principal requirement beyond the normal life of 10 years.

- Optimum inverter topology is needed for ultra-high reliability. It is important to maintain efficiency close to 98 percent for a large central inverter.
- Islanding prevention is mandatory for the grid connection, so that all distribution generators are disconnected in case of grid malfunction.

8.9.5 Utility-Scale Concentrating Solar Power Programs

Solar energy planners are focusing on utility-scale concentrating solar power (CSP) programs with power-generating capacity ranging from 35 to 50 MW. This CSP power system deploys small commercial mirrors and heliostats to minimize system installation cost. The CSP scheme offers more benefits over traditional solar energy-generating technology. The small mirrors used by the CSP plants are capable of tracking the sun within an accuracy of ±0.25 degrees and reflect the solar energy to a tower-mounted receiver sensor, as illustrated in Figure 8.1. The receiver or the boiler converts the water to steam, which powers the turbo-alternator set to generate electricity compatible with grid voltage and frequency requirements. This CSP system makes the renewable energy cost competitive with fossil-based energy-generating systems. A CSP system with generating capacity of 50 MW occupies roughly 160 acres of land and provides a cost-effective solution. A high capacity CSP system can be scaled up to 500 MW or higher using multiple small units to meet the specific power requirements.

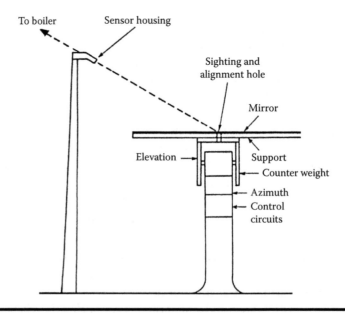

Figure 8.1 A tower-based solar energy collector system.

8.9.5.1 Requirements for Critical Elements and Ideal Locations for CSP Projects

Maximum benefits from the CSP systems require low-cost system elements and ideal locations. The solar energy available from the sun each day is roughly 7 kW-h/m² in a desert environment such as Phoenix, Arizona, or the Sahara Desert in Northern Africa. In colder locations such as Moscow, energy available from the sun is only close to 2 kW-h per day. These examples reveal the ideal locations for the installation and efficient operation of CSP generating systems. The CSP system offers clean, carbon-free electrical energy with lower costs.

Requirements for various system elements should be carefully defined, with particular emphasis on cost and reliability. The system's critical elements include low-cost small mirrors, a tracking mechanism with accuracy better than ±0.25 degrees, a receiver sensor or boiler capable of producing steam and supplying it to the ground-based turbo-alternator module, which is directly coupled to an exciter. The output of the alternator is fed into the grid system through power switches. The steam enters the high-pressure section of the turbine and leaves via the low-pressure section of the turbine to a condenser. The semi-hot condensed water is fed back to the boiler, which receives the concentrated heat energy from the mirrors.

According to the published data [7], typical payback time varies from one year for a location such as Phoenix, Arizona, to four years for a location such as Santiago, Chile. The energy payback time could be two years for other locations such as Johannesburg, South Africa, Madrid, Spain, and Adelaide, Australia. For other locations, such as Moscow, Tokyo, and New York, the payback time could be very close to three years.

8.9.5.2 Solar Thermal Power Systems

The solar thermal power (STP) system is nothing but a miniaturized version of a CSP system. The system design includes a thermal receiver comprising water-filled steel tubes, compact linear Fresnel reflectors, turbo-alternator module, tracking mechanism, and control panel. The control panel monitors randomly the steam pressure, boiler temperature, turbine RPM, alternator voltage and current, and the boiler water level. The compact Fresnel reflectors appear to be solid but actually are made of many smaller, movable reflectors each with a slight curve. These movable reflectors are operating as steerable flat mirrors to achieve optimum thermal performance. Furthermore, the flat mirrors offer lower manufacturing costs compared to parabolic mirrors. In addition, the flat mirrors are closer to the ground and, hence, are less susceptible to wind loads. The system can deploy nearly flat mirrors at ground level that focus the sun's energy into the water-filled steel tubes. The concentrated energy boils the water and converts it into high-pressure, super-heated steam, which drives the turbo-alternator module. The output of the turbo-alternator module is AC electrical energy which can be connected to the utility grid.

An area of 10 km² can accommodate solar collectors enough to generate 700 MW of electrical energy, which can meet the energy requirements of more than 50,000 homes. The cost of electricity generated by an STP system is less than 10 cents per kilowatt-hour. No reliable data on installation costs are available to date.

8.9.5.2.1 Sun Tracking Requirements to Achieve Higher Concentration Ratios

Optimum performance of a PV system is possible, if the irradiation on the solar generator area is continuously vertical. To satisfy this condition, the whole solar generator must be continuously adjusted to the actual position of the sun. Sun tracking can be accomplished in one direction or in two directions. The sun is perpendicular to the solar panel or module only in one plane. The optimum conditions for one-axis tracking are obtained if the panel axis is parallel to the earth's axis. With two-axis tracking the sun is always perpendicular to the solar panel. The PV generator's output could be as high as 5000 W, if the sun is being tracked by a two-axis mechanism. Under this tracking condition, the building is following the elevation of the sun during the day and at night it returns to the position at which the sunrise is to occur. Solar energy collector designers believe that 30 percent higher solar energy can be captured with a two-axis tracking system. In other words, higher concentration ratios are achieved with a two-axis tracking system.

8.9.5.2.2 Concentrating Systems

Low concentration ratio, which varies between 2 and 10, can be achieved through optical components such as mirrors and lenses. When these passive optical devices are used, the solar cell efficiency increases with the logarithm of the sunlight intensity. Note that only direct sunlight can be concentrated. This means that concentrating mechanisms are best suited for very sunny locations. Concentration ratios up to 10:1 are possible with a one-axis tracking system. Standard solar cells can be used for this concentration ratio.

High concentration ratios vary from 10 to 100 suns, while ultra-high concentration ratios range between 100 and 1000 suns. Higher concentration ratios are only possible with a two-axis tracking system. Fresnel lenses are recommended to use in conjunction with a two-axis tracking system, if concentration ratios between 100:1 and 1000:1 are desired. In the case of space applications, each high-efficiency solar cell is equipped with its own concentrator. However, this type of concentration scheme will add additional cost and complexity to the solar power module. High concentration ratios are best suited for boiling water, hot water usage, or generating steam to drive turbine-generator sets.

8.10 Worldwide Photonic Markets and Installation Capacities

This section of the chapter will identify the current solar installation capacities in various countries and will make photovoltaic market projections [8]. In addition, PV power growth over the years in various countries will be identified. In 2007, the global PV market grew by 19 percent with 1750 MW power-generation capacity. Europe continues to lead the world in PV installation. Germany, Spain, and Great Britain allocated huge sums for PV power installations [8].

8.10.1 PV Market Growth in Various Countries

In 2006, Germany added 960 MW of PV power to the grid, which is a 16 percent growth over the previous year. Germany plans to spend $170 million to install solar panels made by the U.S. company First Solar, Inc. near Brandis. This will be the world's largest solar plant and will be operational by the end of 2009. In Germany, the renewable energy industry has become the country's second largest source of new jobs after the automobile sector. Germany employs more than 200,000 scientists and engineers, who are actively engaged in the design, research, and development of solar cells and panels.

Due to lack of fossil fuel sources, Japan started to look toward solar technology. Japanese home owners installed PV power systems about 15 years back and as a result today more than 80 percent of the homes have solar arrays installed. In 1997, the Japanese government started a PV rebate program with $3400 per kilowatt of installed capacity, with a limit of 10 kW installed capacity. However, the rebate was reduced to $2000/kW over the period from 1997 to 2005 and later on it was replaced with a low interest loan. Japan had sales of PV-based products of $3.1 billion, of which about 50 percent of the PV-based components were exported. Today, Japan is the largest PV cell producing country in the world after Germany. A solar cell capacity of 2.4 GW was forecast for 2008, which is close to 350 percent more than 2004. By the end of 2006, 65 percent of the Japanese municipal buildings had already installed PV systems and another 24 percent planned PV installations during 2007 to 2009 for schools, office buildings, street lights, and government offices. Honda Motor Company began mass production in 2006 of CIGS-type solar cells for residential applications. However, such cells are 12 percent efficient and sell for 25 percent less than the silicon solar cells.

The Chinese government has focused on manufacturing solar-heat panels rather than PV cells. China has 150 solar panel manufacturers, producing 33 percent of the world's silicon PV cells. Because the local demand for the silicon PV cells is low, 90 percent of the solar panels produced are exported. A typical PV power system with 5 kW capacity best suited for a medium-sized family costs about $32,000 before rebate and tax incentives, which could be too expensive for a Chinese family. In addition, the cost of electricity generation from solar cells varies from 22 to 28 cents per kilowatt-hour. That is why the Chinese are exporting 90 percent of their solar panels.

As far as power consumption is concerned, Chinese now consumes about 15 percent of the world's energy. But over the last three years that consumption is grown by 70 percent. China's solar power generation capacity is only 25 MW, which is a tiny fraction of the 2830 GW total demand. China's immediate goal is to generate 300 MW of solar energy by 2010. China is producing significant amounts of electrical energy using wind turbines, which typically have an electricity generation cost close to 7 cents per kilowatt-hour. Most of the Chinese electricity needs are met with coal-fired steam-turbo-alternator power plants.

8.10.2 Growth of Solar Installation Capacity

There has been a remarkable worldwide growth of solar installation capacity. The data summarized in Table 8.2 illustrates (within ±5 percent accuracy) the rapid growth of solar power system installation, in various geographical regions and climatic environments. It is evident from the data shown in this table that Germany, Japan, and the United States had roughly 90 percent of the worldwide solar installation capacity in 2005. By the year 2010 that could exceed 94 percent.

These estimates were projected to be significantly higher for the years 2006, 2007, and 2008, because several countries have accelerated the deployment of solar cell technology due to significant increase in the cost of foreign oil.

Despite the significant progress in solar cell technology, several countries around the world are still getting electricity from various fossil fuels, which are much cheaper than solar cells. Various energy sources to generate electricity in the near future are summarized in Table 8.3.

The cost of electricity generation using coal varies between 3 and 6 cents per kilowatt-hour, which is the lowest cost and no other energy source or fossil fuel can cost as little. Solar technology has the highest cost, which currently varies from

Table 8.2 Worldwide Photovoltaic Solar Installation Capacities in 2005

Country	Installed Capacity (MW)	% World Capacity
Germany	1,430	39
Japan	1,425	38
United States	480	13
Spain	74	4
Australia	65	3
Netherlands	56	3

Table 8.3   Potential Energy Sources to Generate Electricity in the Near Future

Energy Source	Percentage of Fossil Fuel
Coal	37
Natural gas	23
Hydroelectric	18
Oil	12
Nuclear	6
Wind energy	2
Solar cell	1.5
Tidal wave	0.5

Table 8.4   Estimated Growth in Solar Energy Generation

Year	Growth in No. of Solar Cells (Billions)	Installed Capacity (GW)
2004	5.5	1.2
2006	12.3	2.2
2008	21.4	3.5
2010	53.8	9.2
2012	65.4	12.5

Note: These estimates are accurate within ±5 percent RMS.

21 to 28 cents, depending on the materials used for fabrication of the solar cells. Nevertheless, according to energy planners, solar power generation is expected to increase at a rapid rate, as illustrated from the data in Table 8.4.

8.11 Performance Capabilities and Cost Estimates for Solar Cells and Panels

This section addresses the performance capabilities and limitations of various solar cells. Fabrication cost for solar cells and installation cost for panels will be discussed in great detail. Fabrication cost of a solar cell is dependent on the material used

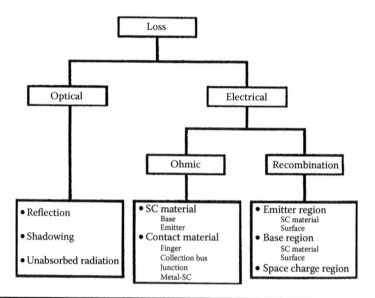

Figure 8.2   Various optical and electrical loss mechanisms in solar cells.

and techniques incorporated to reduce various loss mechanisms such as optical, electrical, ohmic, and recombination losses, as illustrated in Figure 8.2. Cost and efficiency projections for various solar cells are summarized briefly. Estimated panel cost and panel output are identified as a function of panel size and power rating.

8.11.1 *Production Cost and Conversion Efficiency for Various Solar Cells*

Production cost for solar cells is strictly a function of PV material used, number of cells produced in a specific production lot, cell dimensions, and device yield. Conversion efficiency and major benefits of materials used in the fabrication of the first generation of solar cells are summarized in Table 8.5.

Production cost per watt for first-generation solar cells is about $0.85 for organic cells, $1.18 for CdTe cells, $2.70 for thin-film silicon cells, and $3.50 for multicrystalline solar cells. Solar cell designers are predicting 20 to 30 percent lower production costs for the second-generation solar devices. Second-generation thin-film solar cells will have lower costs and higher conversion efficiencies, leading to significant deployment of solar cell technology in the near future. Relative production cost per peak watt and conversion efficiency for various solar cells are shown in Figure 8.3. Cost and efficiency estimates for organic and plastic cells are not provided, because of their extremely low efficiencies, ranging from 3 to 6 percent.

Solar cell cost is a function of cell thickness, device yield, amount of material used, and the dimension of the device. Relative solar cell costs as a function

Table 8.5   Conversion Efficiency and Major Benefits of Various Materials for Cells

Material	Efficiency (%)	Benefits
Gallium arsenide (GaAs)	18–24	Very high efficiency
Multicrystalline silicon	13–16	Moderate efficiency and low cost
Thin-film silicon	14–18	Minimum material required
Cadmium telluride (CdTe)	10–12	Lower cost per watt
Copper-indium gallium selenide (CIGS)	10–12	Lowest cost = $1.18/watt
Organic materials	4–6	Minimum production cost
Plastic	3–5	Very low fabrication cost

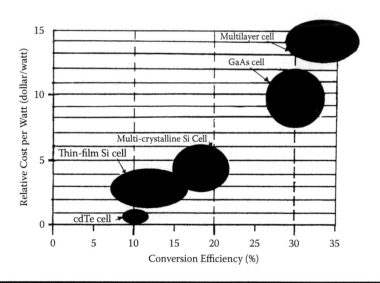

Figure 8.3   Relative energy generation cost per watt and conversion efficiency for various solar cells.

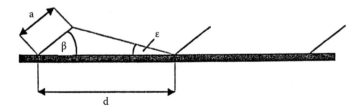

a = Solar module width
d = Spacing between solar arrays
Beta = Tilt angle
Sai = Shading angle

Figure 8.4 Solar panel installation configuration involving a large number of rows of modules to avoid excessive overshadowing. *a*, solar module width; *d*, spacing between solar arrays; β, tilt angle; ε, shading angle.

of cell thickness and cell dimensions are illustrated in Figure 8.4. Evidence of increased cost due to cell breakage and due to use of more material can be seen in this figure.

8.11.2 Solar Panel Cost Estimates and Design Aspects

Housing structure, protective cover, manufacturing tolerances, quality control inspections needed, panel size, and type and number of PV cells installed on a given panel area determine the solar panel production cost. Preliminary solar panel cost estimates received from various panel suppliers and installers indicate that roughly 65 percent of the cost is for solar panels or modules, 15 percent of the cost is for the inverters and the remaining 20 percent of the cost is for the labor required for panel installation. However, the overall solar power system cost for residential installations varies from six to eight dollars per watt of the installed capacity. Based on this cost estimate, a 5-kW solar power system will cost currently anywhere from $30,000 to $ 40,000 before the rebates and tax incentives. According to panel suppliers, both the state and federal tax benefits come to about $10,000, which can have a pay-back time of 10 years for a family with a yearly electric bill of $2000.

Information on panel cost, panel area or size, inverter cost, panel power rating, pay-back time, and number of panels required to meet a specified electrical load have been collected from several solar panel suppliers and are summarized in Table 8.6. These are estimated values only.

According to the panel installers, labor is guaranteed for five years and output power degradation rate is about ½ percent per year. This means after 20 years of operation, the solar system output will decrease by 10 percent. Note the degradation rate increases rapidly, once the operating life exceeds 20 years or so. Since these

Table 8.6   Cost Breakdown for Solar Power System Components and Pay-Back Time

Solar System Component	Panel Supplier and Power Rating		
	A (4 kW)	B (4.5 kW)	C (5 kW)
Solar panel cost	$20,000	$22,000	$24,000
Inverter cost	$3,000	$3,100	$3,200
Electrical and mechanical accessories	$1,000	$1,000	$1,000
Installation cost	$8,000	$8,000	$ 8,200
Total system cost (before tax rebates)	$32,000	$34,100	$36,400
Rebates and tax incentives (approx.)	$15,000	$16,100	$18,000
Estimated net solar system cost	$17,000	$18,000	$18,400
Operating life claimed by manufacturer (years)	25–30	25–30	25–30
Guaranteed inverter life (years)	10	10	10
California utility rebate (approx.)	$7,500	$7,700	$8,200
Federal tax credit (30% of)	Net cost	Net cost	Net cost
Pay-back time (years; based on annual electric bill of $2,000)	8.5	9.0	9.2

cost projections and longevity information are provided by various panel suppliers and manufacturers, the author cannot vouch for the accuracy of the parameters mentioned in the above table.

8.11.3 Pay-Back Period for the System and Performance Degradation Rate for Cells

It is extremely difficult to determine the pay-back period and performance degradation of solar devices, because of several assumptions made, unpredictable operating environments and changes in material properties as a function of age. For a single-family solar power system with 5-kW capacity, the pay-back time ranges from 5 to 10 years depending on the utility monthly bill and the state and federal tax incentives.

As far as the system performance degradation is concerned, the rate of performance degradation is dependent on the operating environment, incident solar energy intensity on the panels, angle of inclination for the panels, tracking mechanism, and the solar cell performance reduction rate as a function of age. Note that most of the incident solar energy on the solar modules or panels goes into purifying the cell materials and encapsulating the solar modules, according to material scientists. In addition, purification of cells is strictly dependent on the incident sunlight that can be absorbed in the shortest time and this varies from location to location. According to solar cell designers, the electrical output of multicrystalline silicon solar cells is reduced by 10 percent after continuous use of 20 years, provided the cells are not subjected to severe mechanical conditions.

In case of organic dye-based solar cells, the rate of performance degradation is very slow and, therefore, the electrical output of such solar cells could last easily for 30 to 40 years. In such solar cells, a mixture of two or more dyes is applied to the glass surface. The dyes absorb the sunlight across a range of wavelengths. The dyes re-emit the light in different wavelengths and transport it across the pane to small solar cells installed at the edges of the plate. This light is collected over a large surface area and concentrated at the edges. This means that the cells need to be located at the edges. This technique not only offers roughly 40 percent more electrical output but also slows the performance degradation of the solar cells. Such solar cells are still in the research and development phase.

8.11.4 Critical Parameters for Solar Panels

The electrical and physical parameters are considered critical for solar panels or modules. The electrical parameters include maximum power output, maximum operating voltage and current levels, open-circuit voltage, short-circuit current, and maximum system voltage. Physical parameters of a typical panel are length, width, thickness, and module weight. Critical parameters from two different suppliers are summarized in Table 8.7.

Panel models A and B are supplied by one vendor, and models C and D by another. Note panels C and D deploy massive integrated technology, eliminating dangling wires, gaps between the panels, and excessive resistive losses. Furthermore panels C and D are longer than panels A and B, but their widths are slightly less.

8.11.5 Sample Calculation for SP-200 Solar Panel

The panel output can be computed provided appropriate or assumed values of relevant panel parameters. The following parameters are assumed.

NTP solar photon intensity: 100 mW/cm^2, which is equal to 92.9 W/ft^2
Panel area: 15.78 sq ft
Absorbing efficiency of the glass cover: 85 percent
Conversion efficiency of the cell: 16 percent
Power input to the solar panel: [92.9 × 15.785 × 0.85] = [1246.96] W
Power output of the solar panel: [0.16 × 1246.96] = [199.5] or [200] W, which is
  the rated output power of panel A shown in Table 8.7.

Table 8.7   Critical Solar Panel Parameters

Parameters	Panel Suppliers and Types			
	A	B	C	D
Cell technology	Multicrystalline	Si	Multicrystalline	Si
Model number	SP-200	SP-170	ST-180	ST-170
Maximum power output (W)	200	170	180	170
Maximum voltage/ current (V/A)	25.4/7.8	34.9/4.8	35.6/5.1	35.2/5.1
Open-circuit voltage (V)	32.3	43.3	44.4	43.8
Short-circuit current (A)	8.40	5.29	5.40	5.41
Maximum voltage (V)	600	600	600	600
Length (in.)	58.03	60.08	62.6	62.6
Width (in.)	39.17	32.51	32.2	32.2
Thickness (in.)	1.4	1.4	2.2	2.2
Weight (lb)	37.0	33.3	8.0	38.0

8.11.6 Electrical Power Consumption Requirements for a Residential Solar System

The design for a residential solar power system requires electrical power consumption needed for all the appliances and computer-based operating systems. The following typical power consumptions must be taken into account prior to installing a specific solar power system.

Central A/C system: 3.52 kW (one-ton capacity) and 7.04 kW (two-ton capacity)
Window A/C unit: 1.76 kW
Latest refrigerator: 0.8 kW
Microwave range: 1.6 kW
Miscellaneous electrical appliances such as toaster, ceiling fan, computer, etc.: 1.0 kW

This means that a solar power system with output capacity close to 5.16 kW is adequate for a medium-sized family of four, provided the family is using a window

A/C unit. It is important to mention that a microwave range or the toaster is seldom used for as long as an hour and, therefore, one can use a washing machine when high-consumption appliances such as the microwave range are not in operation. If the use of a central A/C system is necessary, then one has to increase the solar power system capacity by an additional 3.52 kW for a one-ton capacity unit or 7.04 kW for a two-ton capacity unit.

8.11.7 Typical Performance and Procurement Specifications for Solar Cells and Panels for Residential and Commercial Applications

It is important for anyone who is seriously considering installing a solar power system for a residence or commercial building to get familiar with the performance and procurement specifications for the solar panels or modules. Integrated weather-resistance connections for the panels must be employed for minimum maintenance. Integrated high-performance solar modules must be preferred, because they not only offer a sleek and streamlined structure but also deliver unparalleled reliability of the solar power system using a unique installation with frame structure of high mechanical integrity and weather-protected components. High-efficiency and high-reliability solar panels must be selected to achieve decades of performance with no catastrophic failure.

Published articles on cell materials reveal that about 75 percent of the panel designers are using multicrystalline silicon devices, about 15 percent are using monocrystalline cell technology, and about 10 percent are using thin-film cadmium telluride solar cells. Solar cell material selection must provide high efficiency, improved reliability, and lower fabrication cost, in addition to minimum cell weight.

8.11.7.1 Performance and Procurement Specifications for Solar Cells and Panels Currently Available

The following are the performance and procurement specifications for the solar cells and solar modules currently produced and available on the market.

- Typical cell efficiency: 16 percent (multicrystalline Si cells), 14 percent (monocrystalline Si cells), and 10.5 percent (cadmium telluride cells)
- Maximum panel output power: 170 W, 200 W, or 215 W
- Typical size for the panels currently available: 5 ft × 3 ft
- Typical solar panel area: 15 sq. ft. (minimum)
- Panel longevity or useful life: 25 to 30 years
- Typical 3 ft × 5 ft solar panel weight: 35 (±2) lb
- 3 ft × 5 ft panel cost: $1000 to $1250
- Panel suppliers: First Solar, Solar Home, Nano-solar, Uni-solar, Sanyo, Fuzzi Electric

8.12 Solar Panel Installation Options and Requirements

Panels must be installed on the roof with minimum shadowing effects to achieve near-uniform and maximum electrical energy from the panels. Potential installation options must be evaluated to achieve maximum electrical performance during periods of sunshine. Solar panels can be installed on the surface of a flat roof, on a sloped roof, or on a V-shaped roof structure [9]. Surface-free installation options include racks, integrated into the roof or façade, or mounted at a distance above the building surface to achieve efficient cooling of the solar modules. Note various installation options have their own boundary conditions which will affect the solar system performance and installation cost. Most solar panel installers recommend sloped-roof installation to keep the surface of the panels relatively clean under rainy, dusty, and snow conditions.

8.12.1 Sloped-Roof Installation Option

Studies performed on various installation options reveal that solar panels installed on a sloped roof offer higher reliability, fairly uniform performance, and lower installation costs. In the case of a tiled sloped roof, the solar modules are installed at a distance of a couple of inches from the existing roof tiles. The space between the sonar panels and roof tiles allow cooling of the modules through natural convection. The electrical cable between the solar panel generator and the inverter, which is mostly located inside the house or building, is fed through the roof and must be protected from the adverse effects of rain, heat, and snow. Note the inverter coverts the DC power generated by the solar modules into AC power with appropriate magnitudes of voltage and frequency compatible with the utility grid voltage and frequency values.

8.12.2 Geometrical Considerations for Solar Panel Installation on a Flat Roof

Geometrical considerations play a key role for solar panel installations. Solar panels can be installed on a light-weight support structure on the flat roof's surface. Light-weight mounting frames with high mechanical integrity are readily available on the market. This installation option does not require puncturing of the roof for electrical cable connection between the solar generator and the inverter module because the inverter can be directly located under the solar panels. The inverter must be housed in a weather-proof container for longevity. Installation of the inverter directly on the flat surface requires the shortest connection cable between the solar generator and the inverter. In addition, optimum mounting configuration of solar panels is possible in terms of orientations and inclination. If the solar generators are mounted in rows on a flat roof, a minimum spacing between the rows must be maintained to prevent shadowing effects.

8.12.3 *Impact of Shadowing on Solar Panel Performance*

Optimum boundary conditions are essential for the solar panel installation, improved performance of PV modules, and higher efficiency of the inverter. Note a geographical configuration of a solar installation may involve a large number of rows of solar modules. Appropriate tilting of solar arrays is necessary to minimize the shadowing effects. The optimum tilt angle of large solar arrays containing south-facing modules is approximately equal to the geographical latitude of the installation location, with an error of ±10 degrees. This tilted arrangement offers the best balance of solar energy yield over the year regardless of the seasons. The angle of tilt can be adjusted to achieve optimum system performance in both summer and winter seasons. Note the solar arrays have to be spaced apart by an appropriate distance d with respect to the solar module width a as illustrated in Figure 8.4. The ratio of separation distance to module width can be written as

$$[d/a] = [\cos\beta - \sin\beta/\tan\varepsilon] \tag{8.3}$$

where β is the tilt angle and ε is the shadowing angle of the preceding solar module row and is equal to the sun's azimuth at noon in the winter season. This shadowing angle can be written as

$$[\varepsilon] = [90° - \delta - \varphi] \tag{8.4}$$

where φ is the geographical latitude of the installation location and δ is the ecliptic angle with a value of 23.5 degrees. Computed values of tilt angle (β) as a function of geographical latitude of the installation location and shading angle and for a given ratio of (d/a) are summarized in Table 8.8.

Table 8.8 Computed Values of Tilt Angle as a Function of Latitude and Shading Angle

Geographical Latitude (φ)	Shading Angle (ε)	Tan ε	Tilt Angle β
0 (near equator)	66.5°	2.30	49.0°
10°	56.5°	1.50	40.0°
20°	46.5°	1.05	32.5°
30°	36.5°	0.74	25.0°
40° (near Arctic Circle)	26.5°	0.50	18.0°
50°	16.5°	0.30	11.3°
60°	6.5°	0.11	4.2°

These values of tilt angle are computed for an array spacing of 1 ft and solar module width of 3 ft. It is important to mention that these tilt angles for large arrays compromising south-facing solar modules will yield optimum system performance at noon, regardless of summer or winter season.

According to the solar panel installers, the tilt angle is very flat at locations close to the equator. The shade provided by the solar modules can serve as partial protection against too much sunshine for the panel located underneath. It is important to mention that the tilt angle plays a key role for sun-tracked and concentrating solar systems designed for optimum performance, regardless of the geographical location and time of the day.

8.13 Summary

This chapter summarizes the design aspects and performance capabilities of various energy-generating fossil and nonfossil fuel sources, with particular emphasis on greenhouse effects, reliability, and maintenance requirements to achieve optimum system performance. Alternate energy sources such as coal, natural gas, wind turbine, tidal wave turbines, hydro-turbines, geothermal technology, and utility-scale solar power sources are discussed in detail, with emphasis on reliability, capital investment requirements, and electricity generation cost per kilowatt-hour. Energy sources best suited for defense installations are briefly summarized. Performance capabilities, maintenance requirements, and installation issues for large energy-generating sources such as nuclear reactors, coal-fired steam turbo-alternators, and hydroelectric turbines are discussed in detail. Benefits of microhydro-turbines are highlighted. Performance capabilities, operational issues, and installation requirements for tidal wave and wind turbines are discussed, with emphasis on reliability. Levels of greenhouse gases generated by various energy sources are identified. Solar panel requirements and installation options are discussed in great detail, with emphasis on reliability, longevity, and maximum power output regardless of the seasons and time of the day. Design aspects and operational features of utility-scale concentrating solar power schemes and solar thermal power systems are summarized, with emphasis on concentration ratios, design complexity, and cost of critical components involved in their installation. Details on the worldwide PV market growth and installation capacity are provided for the benefit of the reader and PV system planners and design engineers. Solar panel procurement cost, installation cost, inverter cost, and installation options for solar arrays with minimum shadowing effects are discussed in great detail for the benefit of readers and users of solar energy systems. Higher performance from the roof-installed solar system can be achieved using tracking mechanisms. One-dimensional and two-dimensional tracking mechanisms are identified with emphasis on accuracy of tracking and improvement in system performance regardless of season and time of the day. Critical electrical and physical parameters of solar panels or arrays supplied by various sources are summarized so that the user can select them to meet his or her electrical needs. Solar panel installation options for various roof configurations are described, with emphasis on minimum shadowing effects. Computed values of tilt angle as a function of latitude for the solar system location, shading angle, array separation, and panel width are provided for minimum shadowing effects, regardless of the season and time of the day.

References

1. A. Goetzberger and V.U. Hoffmann. *Photovoltaic Solar Energy Generation.* Berlin: Springer, 2007.
2. A.R. Jha, Technical Proposal and Cost Estimate for a Solar Electric Power System Using Solar Cell Technology. Jha Technical Consulting Services, Cerritos, CA, 1998.
3. Editor. "A less well-oiled war machine," *IEEE Spectrum*, October 2008, 30–32.
4. W.S. Woodward. "Take-back-half: Conversion algorithm stabilizes microhydro-turbine controller," *Electronic Design,* October 2, 2008, 42.
5. Wille P. Jones. "Ocean power catches a wave," *IEEE Spectrum*, July 2008, 14–15.
6. Peter Fairley. "China doubles wind watts," *IEEE Spectrum*, May 2008, 11–12.
7. Warren P. Reynolds. "The solar-hydrogen economy and analysis," *Proceedings of SPIE* 651 (2007): xx.
8. Sandra Upson. "How free is solar energy?" *IEEE Spectrum*, February 2008, 72.

Index

273